JOLPIC SCIENCE!

ORGANIC CHEMISTRY
for Kids

JOLPIC KIDZ

JOLPIC SCIENCE!
ORGANIC CHEMISTRY FOR KIDS

First Published: May 2024

Jolpic Kidz is a publishing company of educational books written for kids.

For more info, mail us at jolpic@gmail.com

Copyright © Jolpic Kidz 2024

All rights reserved. No part of this publication may be reproduced, stored or transmitted in any form, electronic, mechanical, or others without the prior written permission of the copyright owner.

CONTENTS

Let's learn Basic Chemistry — 4

Chemical Bonding — 7

An important concept — 13

What is Organic Chemistry? — 22

Carbon is unique — 24

Structures of Organic Compounds — 27

Hydrocarbons — 32

Functional Groups — 50

Simple Organic Compounds that contain Phenyl Rings — 62

Heterocyclic Compounds — 69

Some Organic Compounds and where to find them — 71

Let's learn
BASIC CHEMISTRY

92 elements exist in nature, and everything around us is made up of these elements.

From the distant galaxies, stars, planets, and asteroids, to the seas, mountains, trees, animals, and all the objects that you can imagine are composed of these 92 elements.

In order to systematically study these elements, scientists constructed the **periodic table** where all the elements are arranged in some rows and columns with respect to their increasing **atomic numbers**.

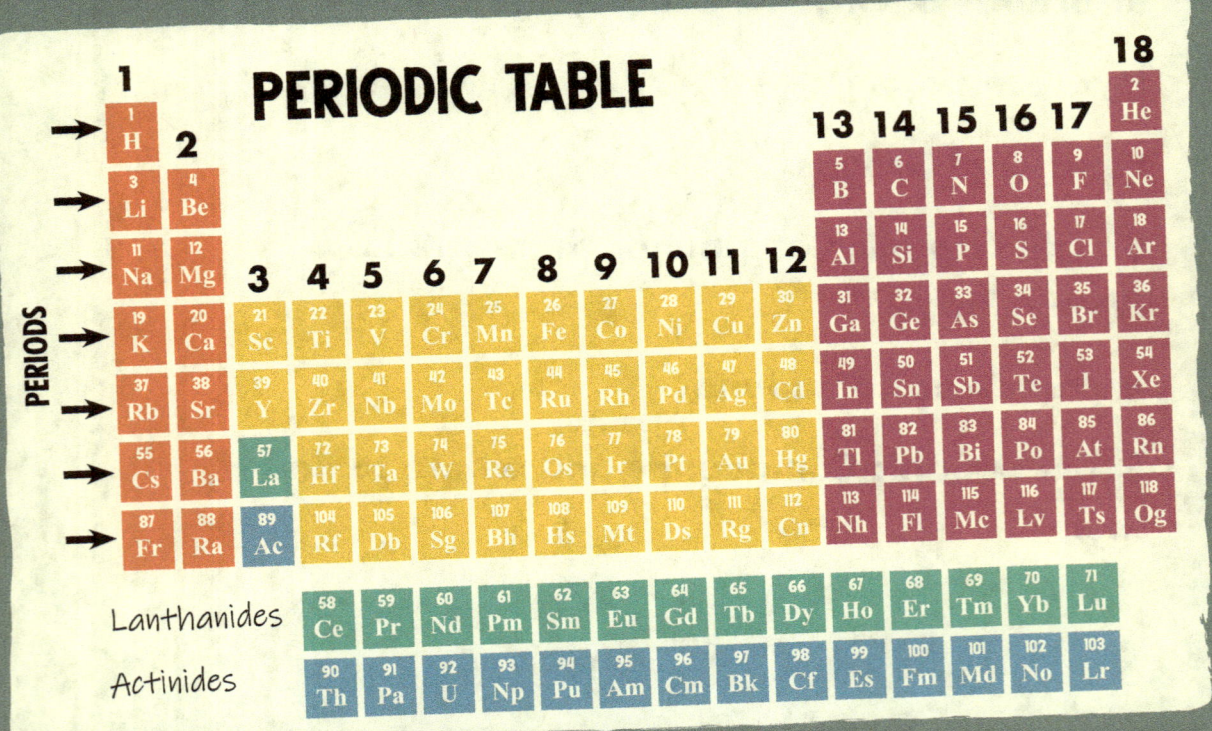

WHAT IS ATOMIC NUMBER?

You probably know that the atom of an element consists of three kinds of subatomic particles, protons, neutrons, and electrons. The protons and the neutrons stay at its nucleus, and the electrons revolve around it.

The number of protons at the nucleus is called the atomic number of the corresponding element.

Different elements have different atomic numbers because the atoms of different elements have different numbers of protons at their nuclei.

For example, the atom of hydrogen has one proton at the nucleus, and one electron is revolving around it. Therefore, the atomic number of hydrogen is 1.

The atom of Helium has two protons and two neutrons at the nucleus, and two electrons are revolving around that nucleus. Since the number of protons at the nucleus is 2, the atomic number of helium is 2.

If you look at the periodic table, you see the elements are written in their short forms. For example, hydrogen is written as H, helium is written as He, and so on.

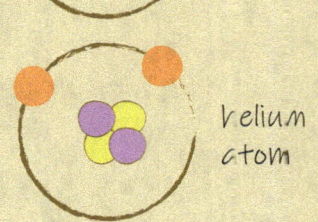

These short forms of elements are known as symbols.

Now you may ask one possible question:

"If there are only 92 naturally occurring elements, why do we see numerous substances around us?"

Some of these 92 elements combine themselves in various proportions to form compounds. The combined forms of atoms are called molecules.

> You know that there are 92 naturally occurring elements, but the total number of elements in the periodic table is 118. How is this possible?
>
> It is possible because 118 - 92 = 26 elements are created by scientists in the laboratory. These 26 elements are not found in nature.

For example, the compound called water that we drink everyday is formed by the combination of elements, hydrogen and oxygen.

Two atoms of hydrogen combine with one atom of oxygen to form a water molecule.

Carbon dioxide is a gaseous compound, and each molecule of carbon dioxide consists of one atom of carbon and two atoms of oxygen.

Numerous numbers of such compounds are available to us.

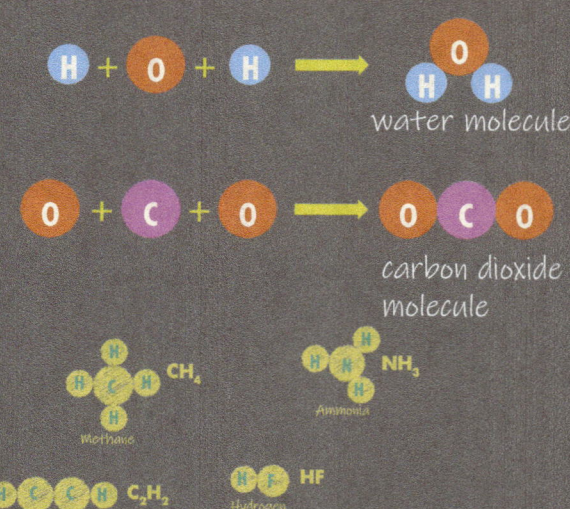

The atoms of many of the elements do not prefer to stay in their mono-atomic states. Therefore, they exist sometimes in their diatomic, triatomic, or polyatomic states in their elemental forms.

CHEMICAL BONDING

A chemical bond acts as the adhesive between atoms, so that two or more atoms join together to form a molecule. The electrons, which are revolving around the nucleus of an atom, play a major role in forming chemical bonds.

Two types of chemical bonds are generally found in different molecules.

Ionic Chemical Bond

Covalent Chemical Bond

Atom Atom

Bonded Atoms

Apart from ionic bonding and covalent bonding, there are a few more types of bonding in chemistry, such as coordinate bonding, hydrogen bonding, etc. However, we are not going to talk about them in this book.

HOW IONIC BONDS ARE FORMED

You know that an atom is made up of three kinds of subatomic particles, neutrons, protons, and electrons. Among these three, protons are positively charged, electrons are negatively charged, and neutrons are neutral or chargeless.

You also know that the nucleus of an atom of a particular element consists of a definite number of positively charged protons and chargeless neutrons. Therefore in the case of a neutral atom, the number of protons and the revolving electrons should be equal.

For example, the nucleus of a sodium atom consists of 12 neutrons and 11 protons. Since we are talking about a neutral sodium atom, the number of electrons in its obits is also 11.

Similarly, a neutral chlorine atom also can be shown. The nucleus of a chlorine atom consists of 18 neutrons and 17 positively charged protons. Thus the number of electrons in the neutral chlorine atom is 17.

One important fact is that the electrons can not revolve in any possible orbit around the nucleus of an atom, but in some specific orbits, and each orbit is able to hold a specific number of electrons.

For example, maximum 2 electrons can stay in the first orbit, maximum 8 electrons can stay in the second orbit, and so on.

Therefore the rough pictures of sodium and chlorine atoms will look like,

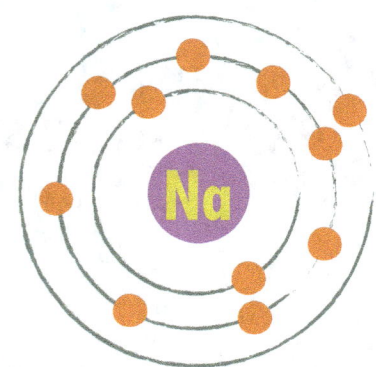

For sodium atom,
First orbit holds 2 electrons
Second orbit holds 8 electrons
Third orbit holds 1 electron

Total 11 electrons

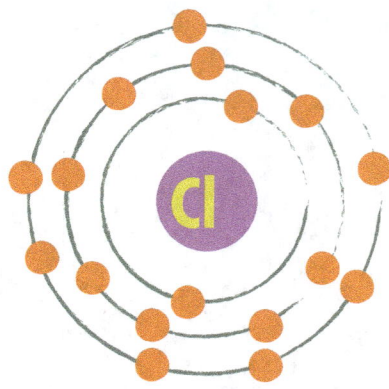

For chlorine atom,
First orbit holds 2 electrons
Second orbit holds 8 electrons
Third orbit holds 7 electrons

Total 17 electrons

The ionic bonding is feasible between sodium and chlorine. Let's understand how this ionic bond forms.

One electron transfers from the outermost orbit of a sodium atom to the outermost orbit of a chlorine atom.

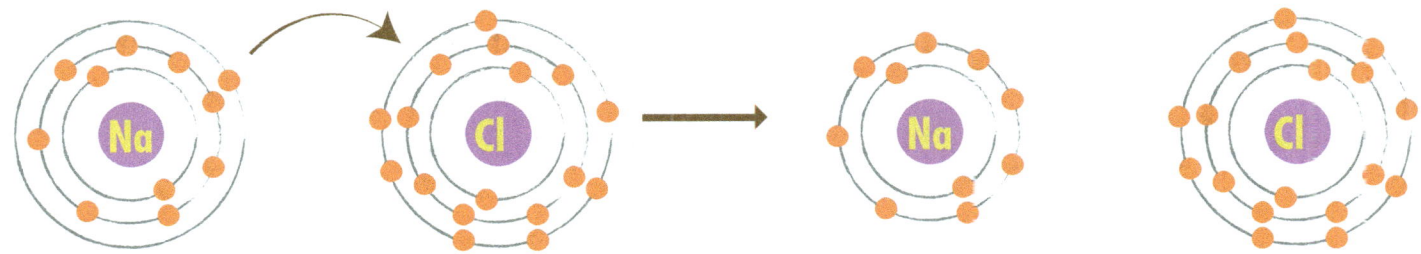

Notice that, after the transfer of one electron, the number of protons and the number of electrons do not remain equal for each type of atom.

 Number of protons = 11
Number of electrons = 10

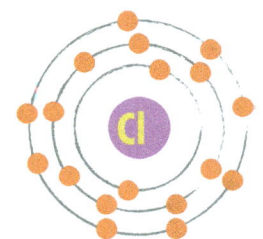 Number of protons = 17
Number of electrons = 18

Due to the presence of one extra proton at the nucleus of sodium, the overall charge becomes positive. The positively charged sodium is now called a sodium ion or a sodium cation.
Likewise, the chlorine atom now has one extra electron in its orbit. Therefore, the overall charge becomes negative. It is now called a chloride ion or a chloride anion.

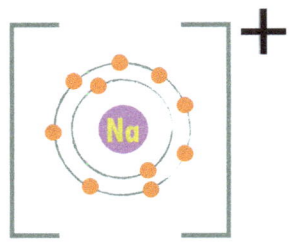
Sodium cation

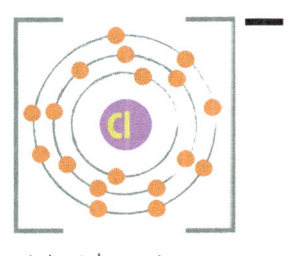
Chloride anion

An ion is formed when an atom or a group of atoms possesses unequal numbers of protons and electrons. A positively charged species is called a cation, and a negatively charged species is called an anion.

Probably you know that unlike charges always attract each other due to electrostatic interaction. The same will happen in the present scenario. The positively charged sodium cation will attract the negatively charged chloride anion, and they will combine.

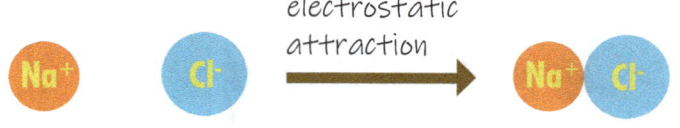

When oppositely charged ions stick together due to electrostatic force, the bonding is called an ionic bonding.

The above discussion shows how the compound called sodium chloride is formed due to the attractive interaction of oppositely charged ions. Since here the bonding between sodium and chlorine is an ionic bond, sodium chloride is called an ionic compound.

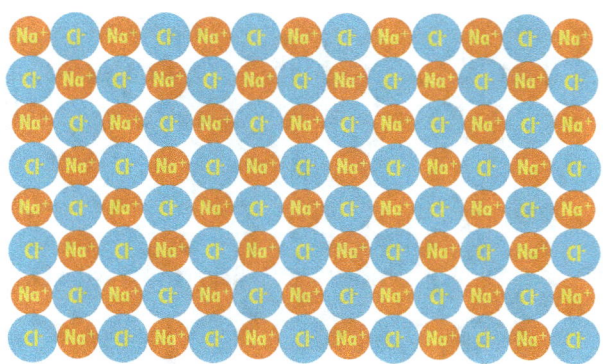

How a large number of sodium and chloride ions stay arranged in sodium chloride compound.

We intake this compound everyday with our food, and this compound is commonly known as table salt.

There are many other examples of ionic compounds.

Sodium chloride

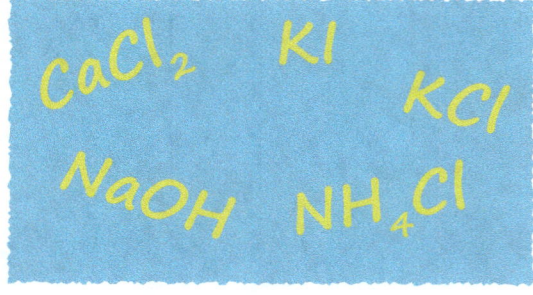

$CaCl_2$ KI KCl $NaOH$ NH_4Cl

HOW COVALENT BONDS ARE FORMED

An ionic bond is formed when the complete transfer of electron/s occurs between two atoms, and they stick together by electrostatic force. However, a chemical bond can also be formed by sharing electrons. Such chemical bonds are called covalent bonds.

Let's understand covalent bonding with a few examples.

A hydrogen atom has one proton and no neutron at the nucleus. Since a free hydrogen atom has to be neutral (chargeless), it has one electron revolving around the nucleus.

Hydrogen Atom

But in nature, the elemental hydrogen is always diatomic. Two hydrogen atoms join together to form a hydrogen molecule.

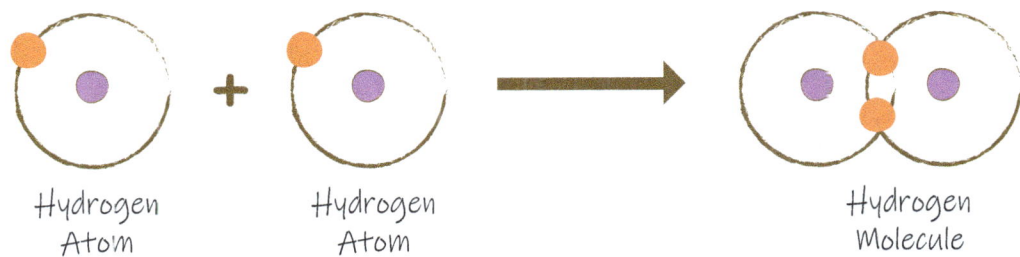

In the hydrogen molecule, the bond between two hydrogen atoms is a covalent bond, which is associated with electron sharing.

Consider two hydrogen atoms, hydrogen A and hydrogen B, are going to make a covalent bond between them. Before the bond formation, the electron on each hydrogen atom belongs to that atom only.

Before the bond formation

But, when electrons are shared to form a covalent bond, each electron belongs to both of the hydrogen atoms. You cannot tell which electron comes from which hydrogen atom simply looking at a hydrogen molecule.

After the bond formation

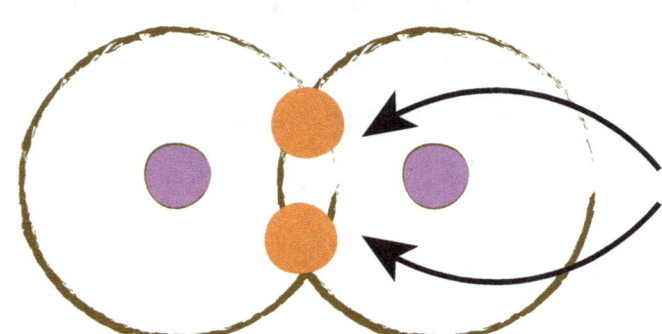

After the covalent bond formation, both the electrons now belong to atom A and atom B. It means that you can equally find each of the electrons around the nuclei of atom A and atom B.

The covalent bond between two hydrogen atoms is associated with one shared pair of electrons, so this bond is a single covalent bond.

This single bond can be represented as simply drawing a straight line between two hydrogen atoms.

$$H\!-\!H$$

When covalent bonds are formed by sharing two or three pairs of electrons, they are called double bonds or triple bonds respectively.

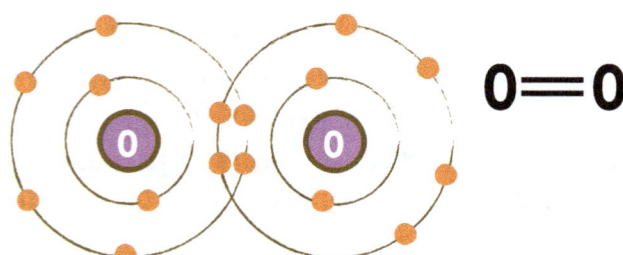

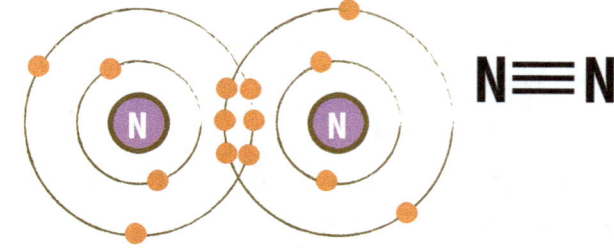

The covalent bond between two oxygen atoms is associated with two shared pairs of electrons, so it is a double bond.

Similarly, the covalent bonding between two nitrogen atoms is associated with three shared pairs of electrons. Thus the covalent bond between two nitrogen atoms is a triple bond.

An Important Concept

A particular element generally makes a specific number of covalent bonds with other elements. It means the number of shared electrons is fixed for a particular element. The number of electrons in the outermost shell of a particular atom determines the number of electrons that will be shared with other atoms.

It was discussed previously that each shell of an atom can accommodate a particular number of electrons.

The first shell accommodates 2 electrons. Thus there will be 2 electrons in total after electron sharing to make the covalent bond, which is associated with the first shell. Similarly, the second shell accommodates 8 electrons. Therefore the total number of electrons after the sharing will be 8 when the outermost shell is the second shell.

Hydrogen Atom

The hydrogen atom has 1 electron in its orbit, and thus it needs one more electron in order to achieve its closed shell configuration. Therefore, the atom of hydrogen is allowed to form one covalent bond sharing one electron pair with another atom.

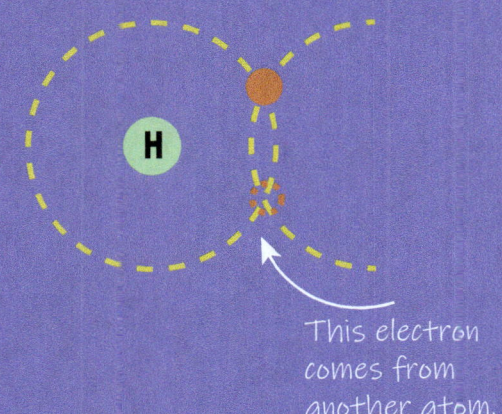

This electron comes from another atom.

The hydrogen atom is able to make one covalent bond with other atom.

Carbon Atom

The carbon atom has following electronic structure,

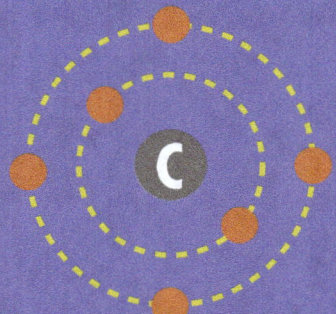

Maximum 8 electrons can be accommodated in the second orbit, so 4 more electrons are required to attain its closed shell configuration. It is possible when a carbon atom makes 4 covalent bonds sharing four electron pairs with other atoms.

Therefore, a carbon atom makes 4 covalent bonds.

Nitrogen Atom

Nitrogen atom has 5 electrons in its outermost orbit.

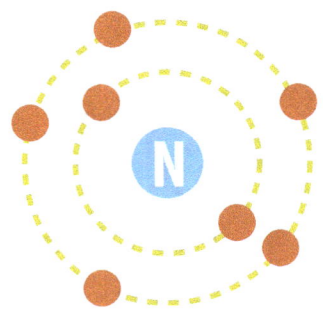

When the nitrogen atom makes 3 covalent bonds sharing 3 pairs of electrons with other atom/s, the closed shell configuration of 8 electrons will be attained.

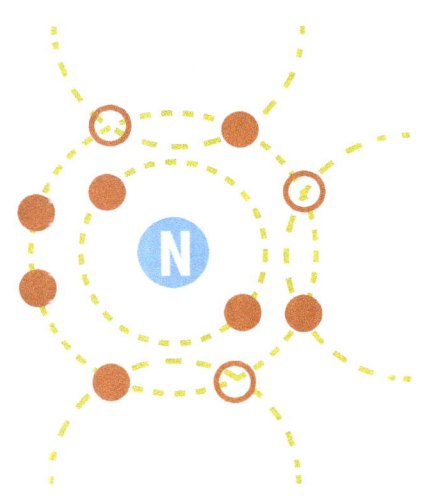

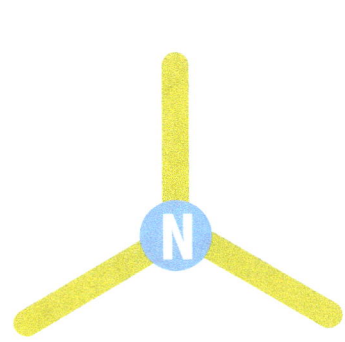

Therefore, the nitrogen atom generally makes 3 covalent bonds with other atoms.

Oxygen Atom

There are 6 electrons in the outermost orbit of the oxygen atom. When 2 pairs of electrons are shared with other atoms, oxygen attains the closed shell configuration.

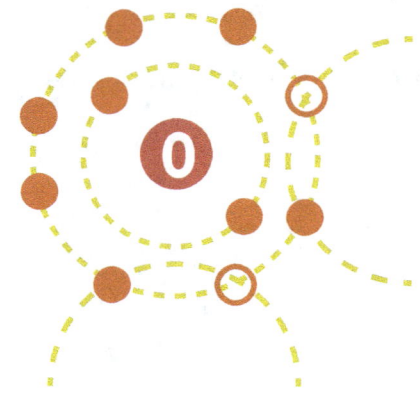

Therefore, oxygen makes two covalent bonds with other atoms.

Fluorine, Chlorine, Bromine, Iodine

These elements generally make one covalent bond with another atom.

You may imagine the following pictures in your mind to show the number of covalent bonds a particular atom can form.

Hydrogen atom has one hand.

Carbon atom has four hands.

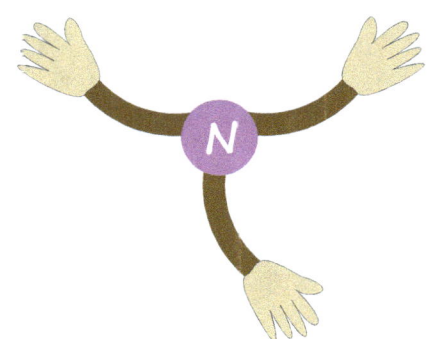

Nitrogen atom has three hands.

Oxygen atom has two hands.

Fluorine atom has one hand.

Chlorine, bromine, iodine also have one hand for each.

Now we are able to determine the number of covalent bonds between atoms in various molecules.

Water molecule

The water molecule consists of two hydrogen atoms and one oxygen atom.

Hydrogen Atom   Hydrogen Atom   Oxygen Atom

The oxygen atom has two hands and the hydrogen atom has one hand. Therefore, the oxygen atom will hold the hands of two hydrogen atoms with its hands.

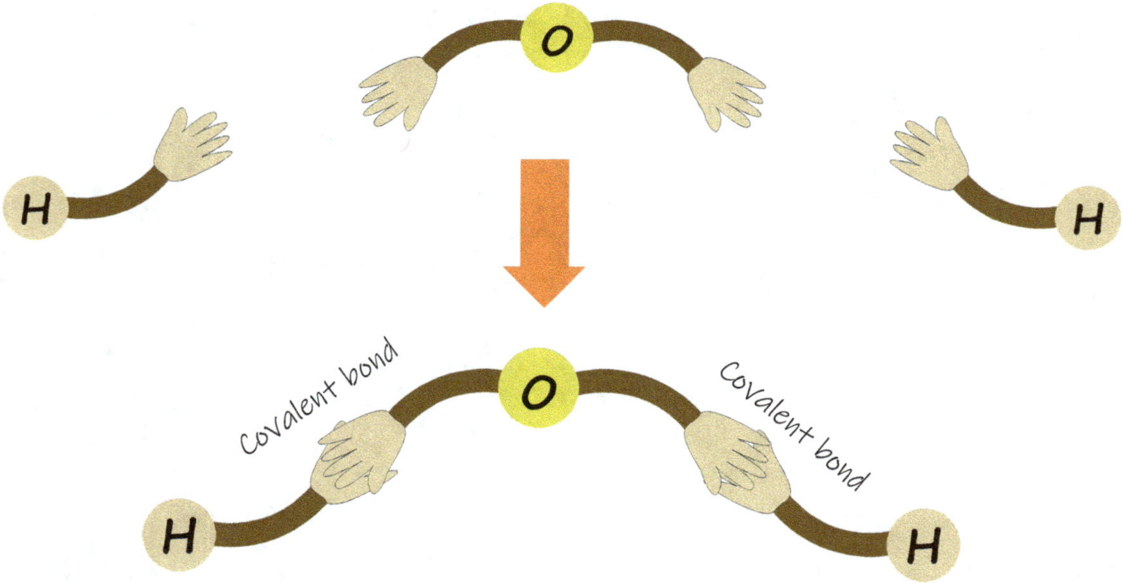

Each pair of holding hands represents a covalent bond. Therefore, the water molecule has two hydrogen-oxygen covalent bonds.

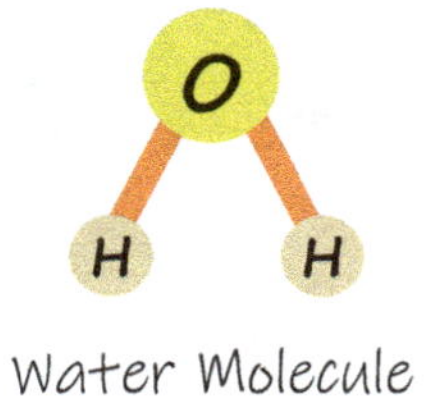

Water Molecule

Ammonia molecule

The ammonia molecule consists of one nitrogen atom and three hydrogen atoms.

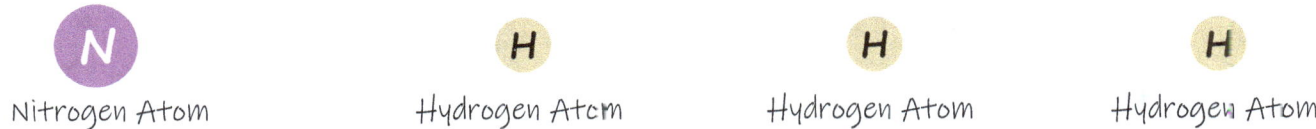

The three hands of the nitrogen atom will hold the hands of three hydrogens in order to form an ammonia molecule.

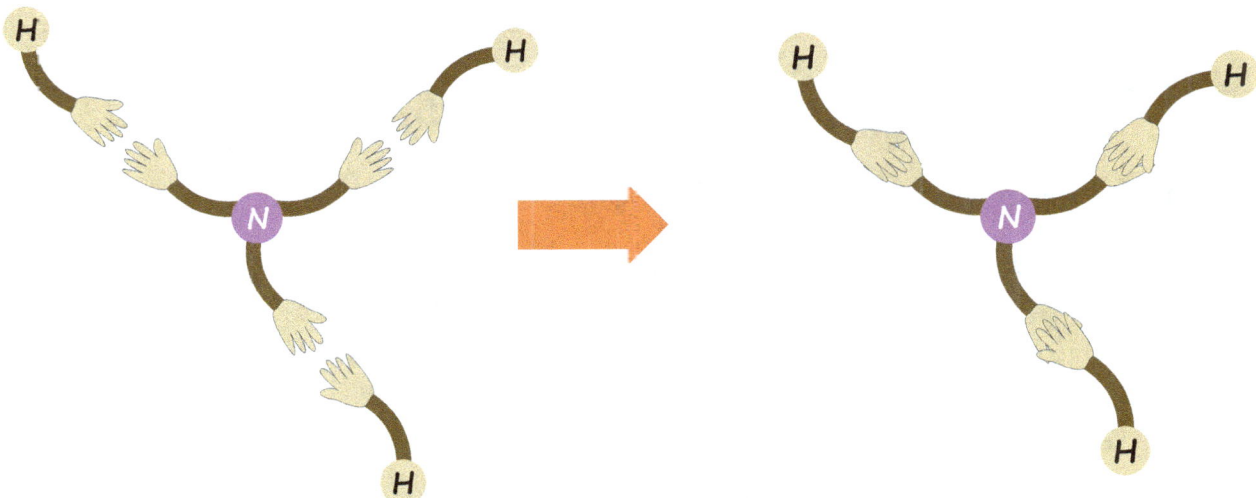

Therefore, there are three N-H covalent bonds in the ammonia molecule.

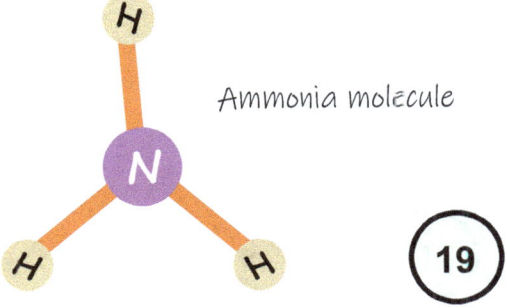

Ammonia molecule

Hydrogen fluoride molecule

The hydrogen fluoride molecule consists of one hydrogen and one fluorine atom.

Each of these atoms possesses one hand. Therefore, they will utilize their hands to hold each other.

The hydrogen fluoride molecule has one H-F bond.

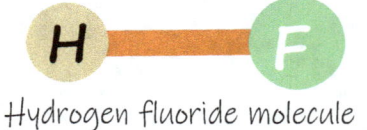

Hydrogen fluoride molecule

Methane molecule

The methane molecule consists of one carbon and four hydrogen atoms.

The carbon atom has four hands available to hold four hydrogens. Therefore, the methane molecule has 4 C-H covalent bonds.

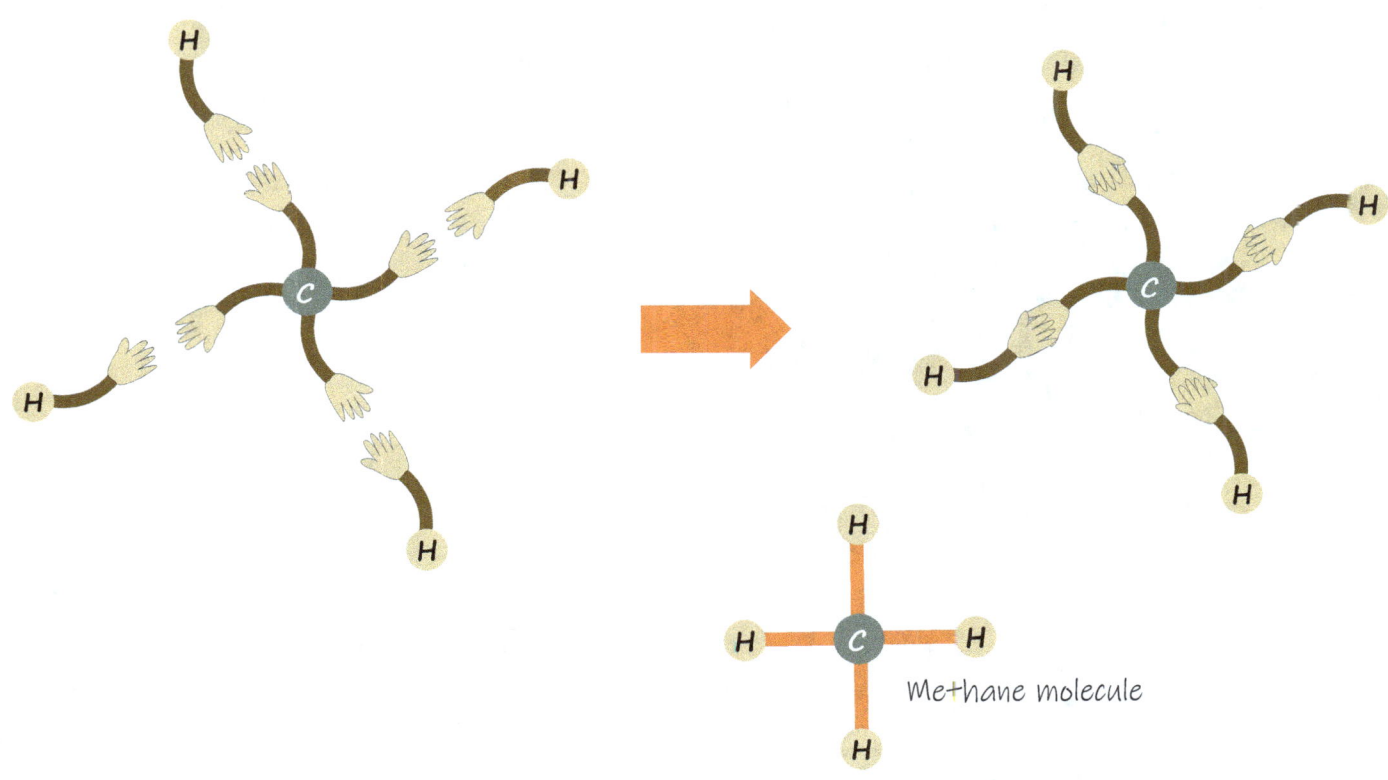

Methane molecule

All these discussions about basic chemistry are not sufficient to tell the atomic structures of all the elements, or the bonding features of all the compounds that are known to us today. However, the discussions are quite sufficient to teach basic organic chemistry to young people like you.

What is Organic Chemistry?

You may guess from the term 'organic' that the subject is somehow related to organisms. You guessed right. All the organisms that live on Earth are composed of a special class of compounds called organic compounds.

As you have a little knowledge of biology, you know that every organism is made of cells. Some organisms are unicellular, and others are multicellular, but the chemical compositions of all those cells are organic compounds.

Therefore everything that we get from animals and plants is primarily a mixture of organic compounds.

Our foods are derived from plants and animals, thus we intake millions of organic molecules everyday.

Vinegar, the ingredient we often use during the cooking of food, is also a water solution of an organic compound known as acetic acid. Camphor is an organic compound obtained from a kind of plant.

This book is made of paper, and paper is made from wood. Therefore, his book also contains organic compounds primarily cellulose. Our urine contains an organic compound called urea.

However, it does not mean that all organic compounds are necessarily to be found in organisms. Many organic compounds have been synthesized in the laboratories also. For example, there are several medicines that are manufactured by pharmaceutical companies in the laboratories.

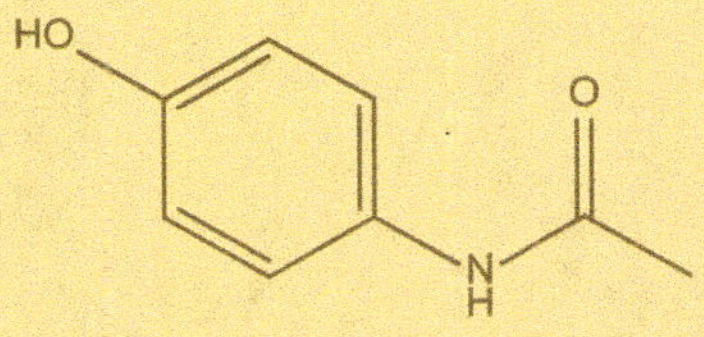

Paracetamol

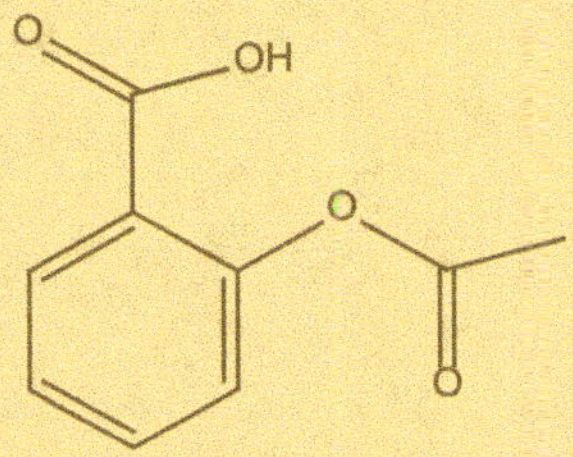

Aspirin

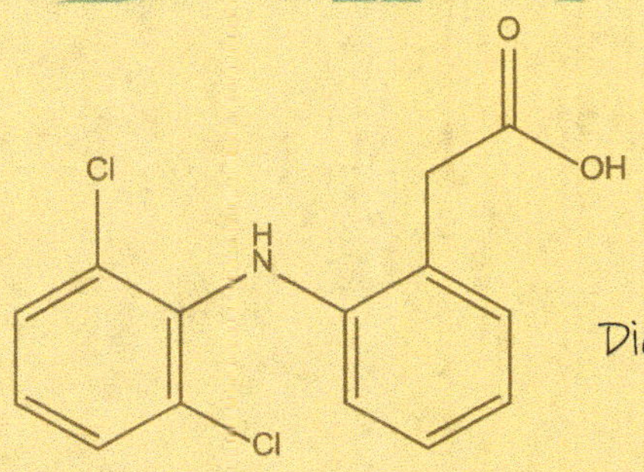

Diclofenac

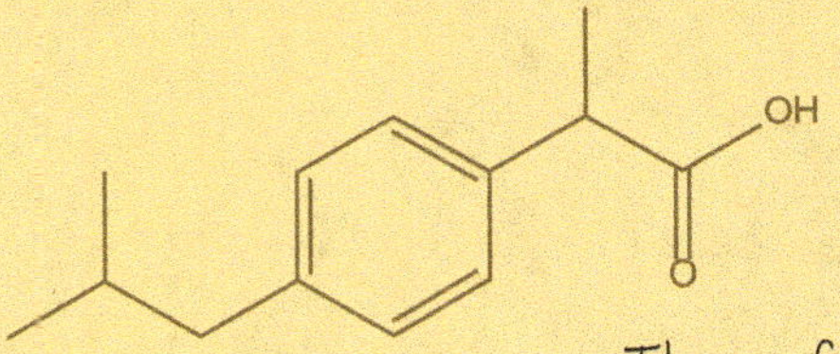

Ibuprofen

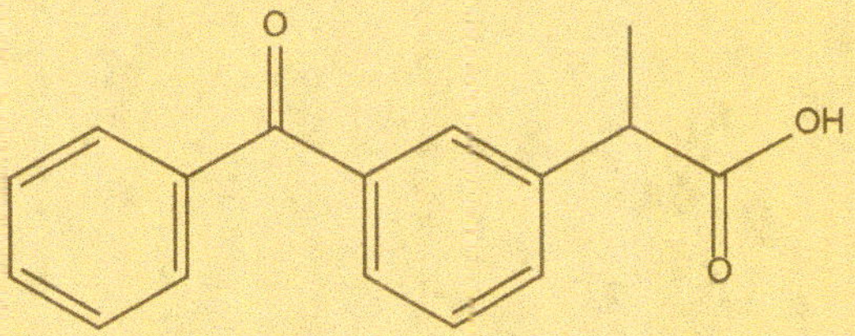

Ke-oprofen

> **" The important fact about any organic compound is that its molecular framework must contain carbon atoms. "**

Since organic compounds are the building blocks of all organisms, and the carbon atom is the primary constituent of organic molecules, it may be concluded that no organism would exist on Earth if the carbon atom were absent in nature.

Carbon is Unique

Carbon is quite unique among all 118 elements in the periodic table, because carbon is able to form millions of organic compounds combining with hydrogen, oxygen, nitrogen, etc. No other element in the periodic table possesses the capability to form such a huge number of compounds.

At this point, you need to recall what we previously discussed about carbon. The nucleus of a carbon atom consists of 6 positively charged protons and 6 chargeless neutrons. Therefore a neutral carbon atom has 6 electrons revolving in some specific orbits around the nucleus.

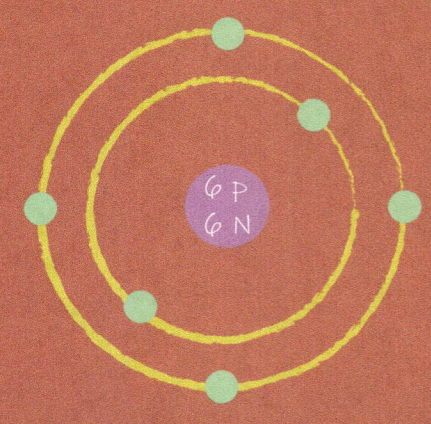

You know that maximum 2 electrons can stay in the first orbit, and maximum 8 electrons can stay in the second orbit. Thus we may draw a rough picture of a neutral carbon atom as,

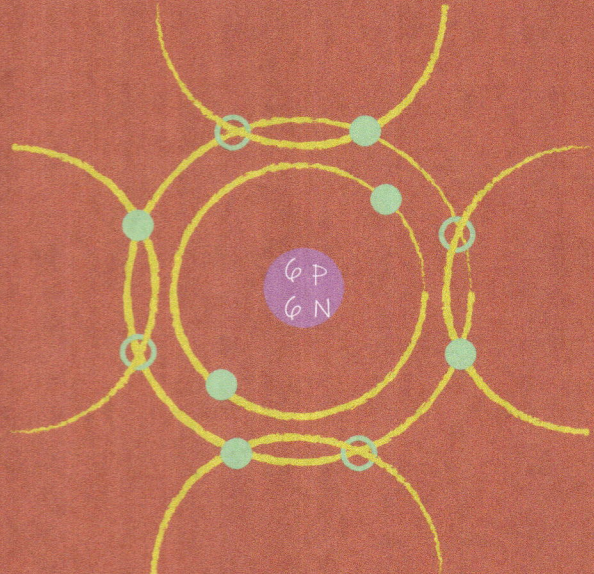

We also discussed how the carbon atom attains its closed shell configuration through formation of covalent bonds by sharing electrons. Since the outermost orbit has 4 electrons, it will need 4 more electrons to achieve the closed shell configuration. Thus when the carbon atom makes four covalent bonds with other atoms by sharing electrons, the total number of electrons in the outermost shell becomes 8.

Generally, the carbon atom makes four covalent bonds with other atoms. It can be also imagined as the carbon atom has four hands to hold other atoms.

A carbon atom may use these hands to hold other carbon atoms also.

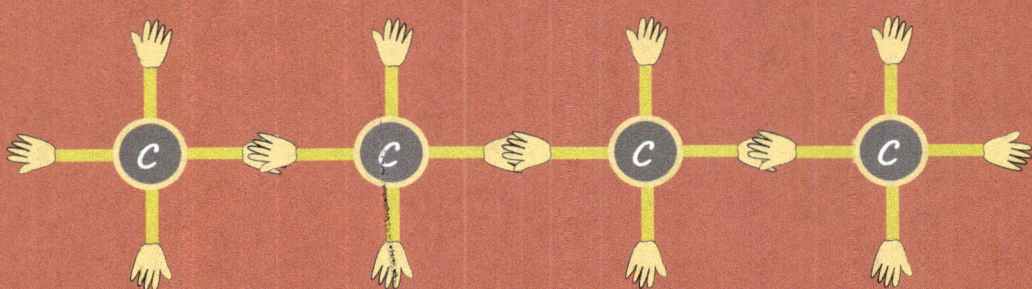

It means that carbon can make covalent bonds among themselves without any difficulty, and that way, several carbon atoms are able to make long or short carbon chains or branches through covalent bonds.

This particular property of carbon is known as catenation. No other element in the periodic table is able to show such an efficient catenation property like carbon.

A given number of carbon atoms may bond together in several ways. And the remaining available positions for bonding can be satisfied by either making multiple bonds or making covalent bonds with other atoms or groups of atoms.

For example, all these organic compounds contain four carbon atoms.

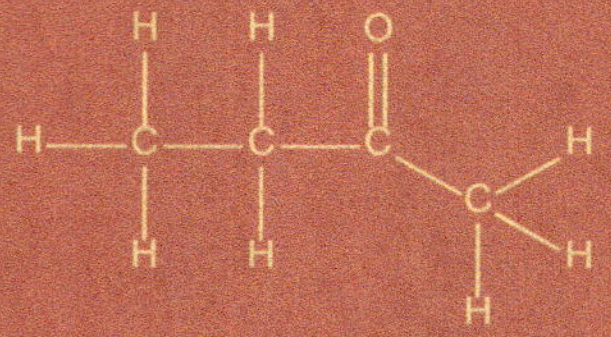

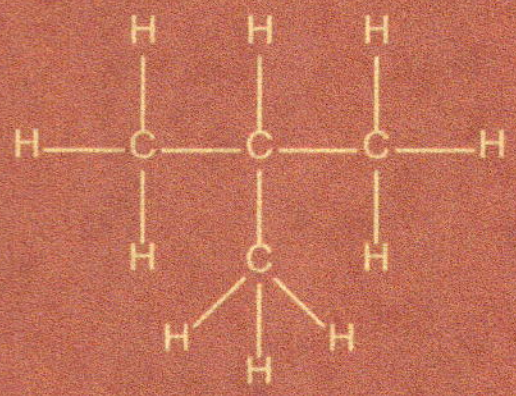

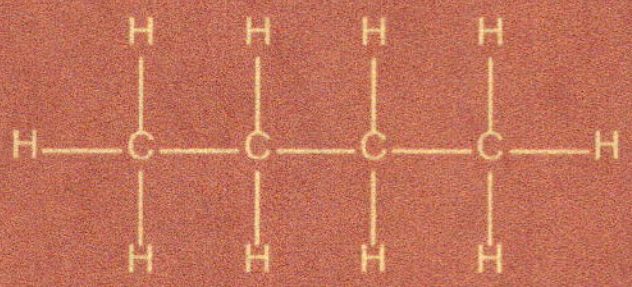

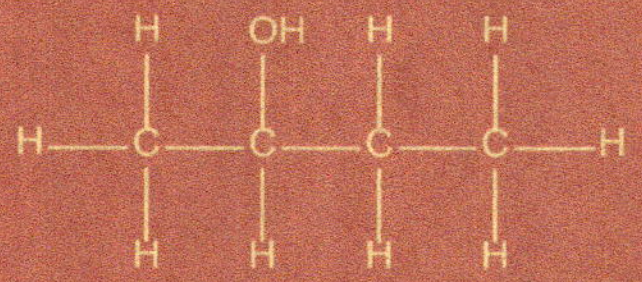

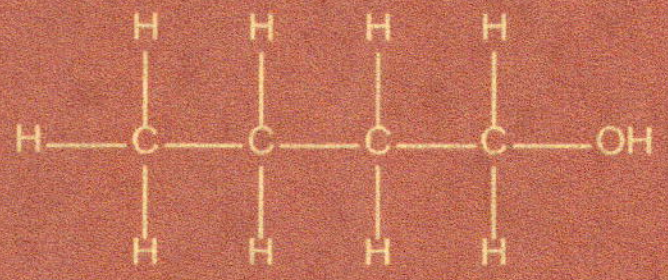

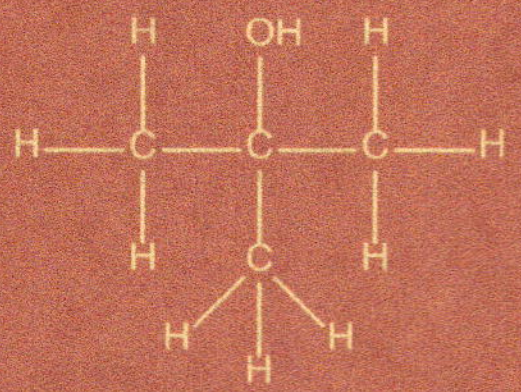

The list does not end here. Several other organic compounds are possible just with four carbon atoms.

Now imagine how many organic compounds are possible when you increase the number of carbon atoms. The number of compounds will be millions.

Hope you now realize why millions of organic compounds exist in nature.

The study of structures, chemical properties, reactions, and synthesis of organic molecules is simply known as organic chemistry.

Structures of Organic Compounds

From the previous discussions, you have learned that a carbon atom can make 4 covalent bonds, and a hydrogen atom can make 1 covalent bond.

For simplicity, you may imagine the situation as the hydrogen atom has one hand, and the carbon atom has four hands. When covalent bonds will be formed between them, the atoms will hold the hands of each other.

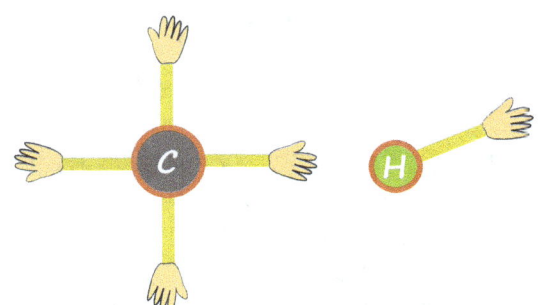

Now we may construct the structure of the simplest organic molecule, which consists of one carbon atom and four hydrogen atoms.

The carbon atom holds the four individual hands of four hydrogen atoms with its four hands.

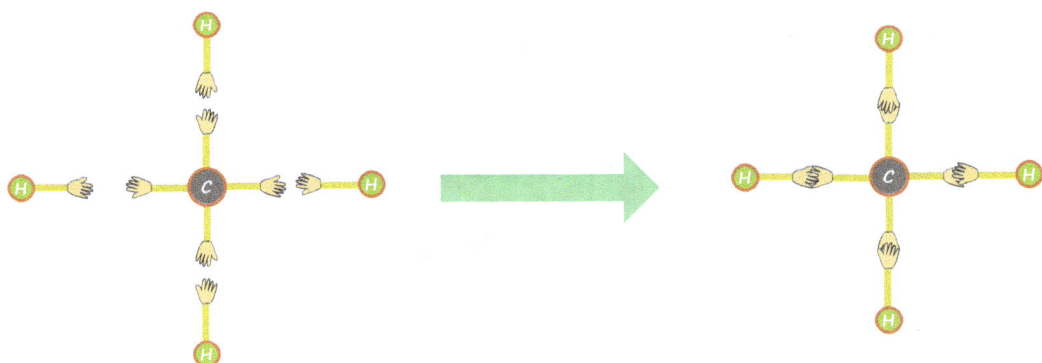

It means the carbon atom forms four C-H bonds with four hydrogen atoms.

This organic compound is known as methane.

The methane molecule consists of carbon and hydrogen atoms only, so it is a hydrocarbon.

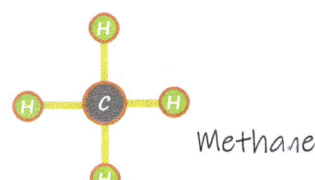
Methane

When an organic molecule is made up of carbon and hydrogen, and no other element is present in it, the compound is called a hydrocarbon.

When a carbon atom makes four single covalent bonds with other atoms, those four bonds are pointed along certain directions in space.

In the methane molecule, the four C-H bonds are directed along the vertices of a regular tetrahedron.

Place the carbon atom at the center of a regular tetrahedron, and draw four hydrogen atoms at the four vertices. Then connect the four hydrogen atoms to the carbon atom drawing four covalent bonds.

Now you can see the adjacent C-H bonds are spatially separated at certain angles. If you measure the angle between any adjacent pair of C-H bonds, you will find its value, 109.47 degree.

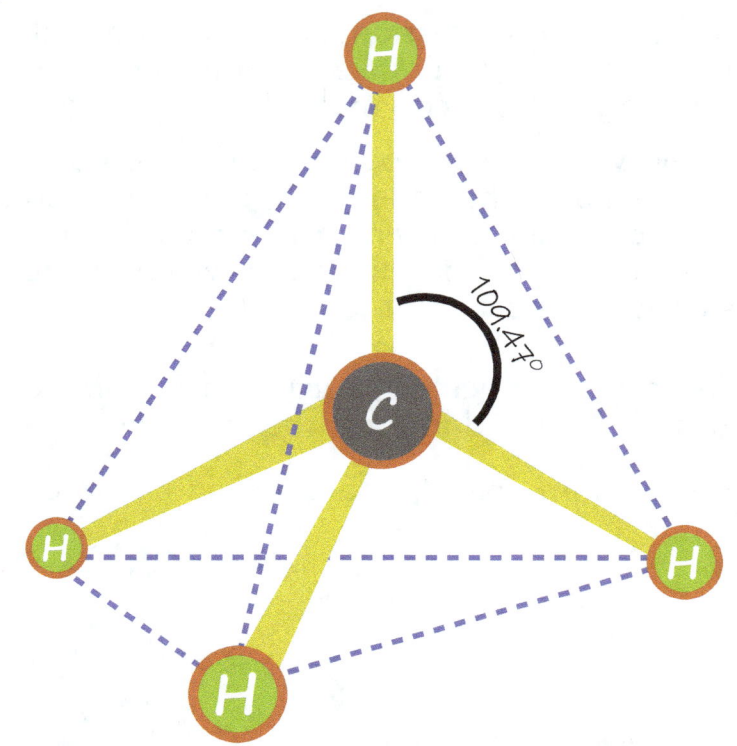

For that reason, organic molecules are often represented by wedge-dash notation.

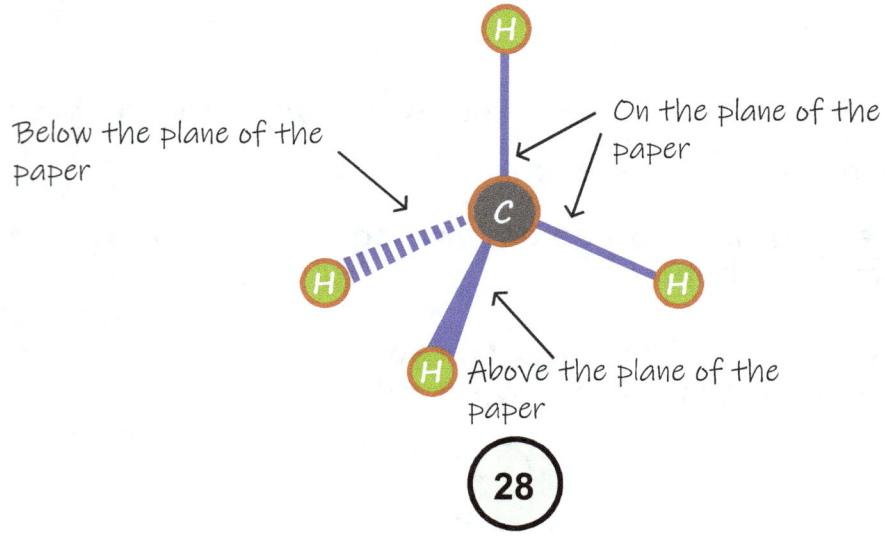

Many organic molecules often contain double bonds and triple bonds.

Let's construct the structure of an organic molecule consisting of two carbon atoms and four hydrogen atoms.

First make a covalent bond between two carbon atoms.

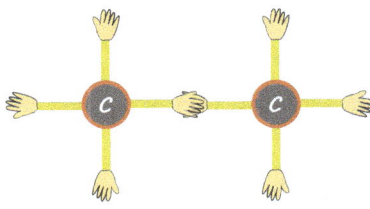

Then the four hydrogen atoms will make single bonds holding the available hands of the carbon atoms.

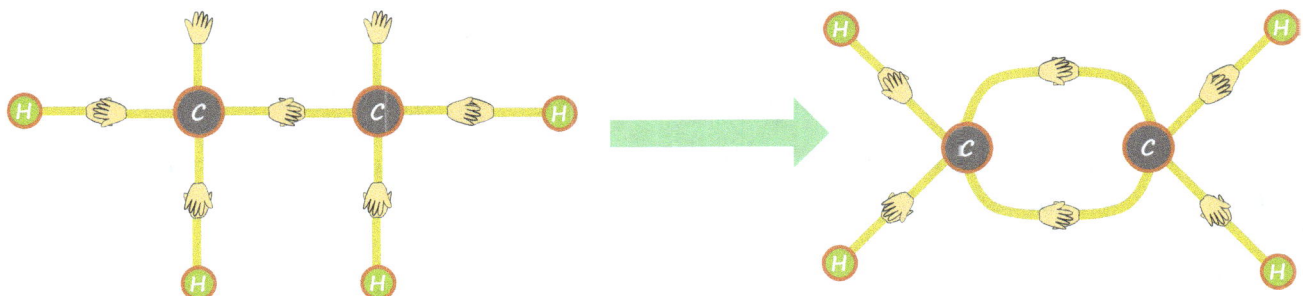

Notice, the carbon atoms still have a pair of available hands for bonding. The hands will hold each other to make another covalent bond between two carbon atoms.

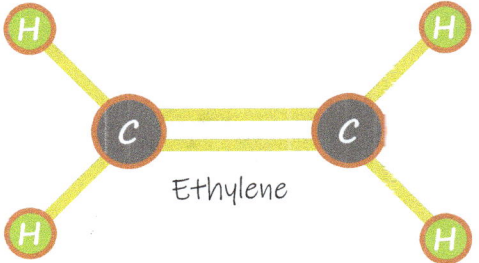

Ethylene

Therefore, the covalent bond between the two carbon atoms is a double bond.

Similarly, carbon-carbon triple bonds are also found in organic molecules.

Let's try to draw the structure of a molecule consisting of two carbon and two hydrogen atoms.

First we draw a carbon-carbon single bond between two carbon atoms.

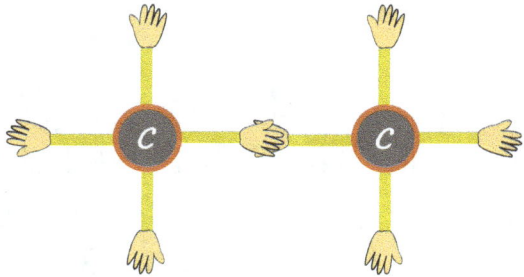

Then two carbon atoms make covalent bonds with two hydrogen atoms.

The remaining available hands on the carbon atoms will make two more covalent bonds between the carbon atoms.

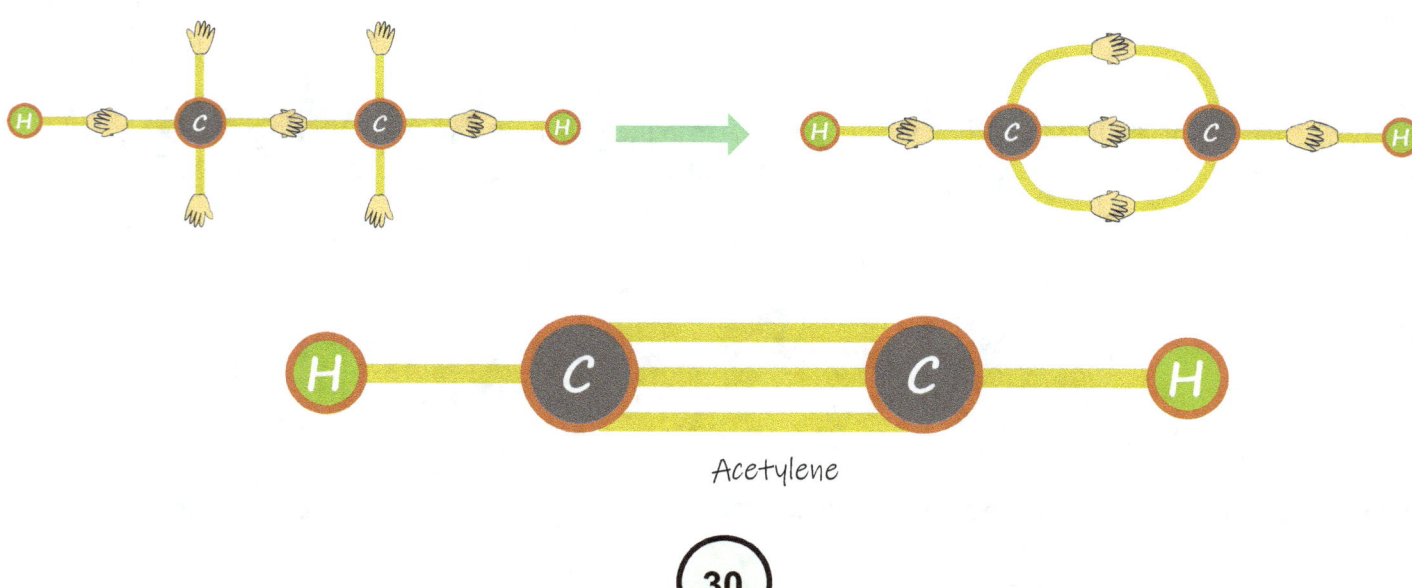

Acetylene

In any stable organic molecule, each carbon atom is susceptible to form 4 covalent bonds. The formation of these four covalent bonds may take place in several ways.

These bonds may be,

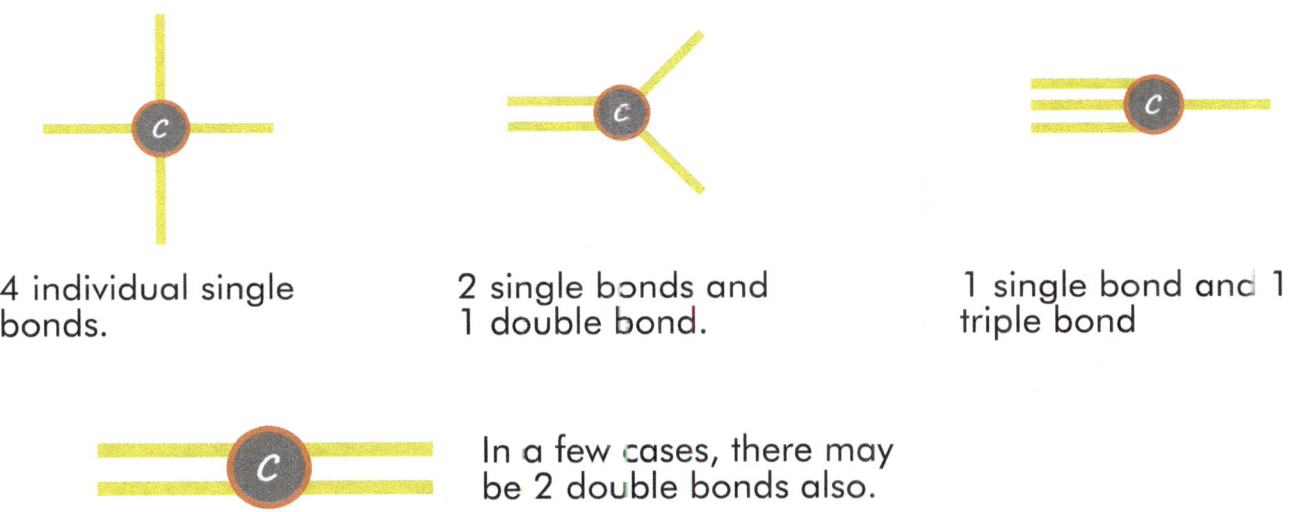

4 individual single bonds.

2 single bonds and 1 double bond.

1 single bond and 1 triple bond

In a few cases, there may be 2 double bonds also.

The angle between the bonds varies from 109.47 degree when a multiple bond is concerned.

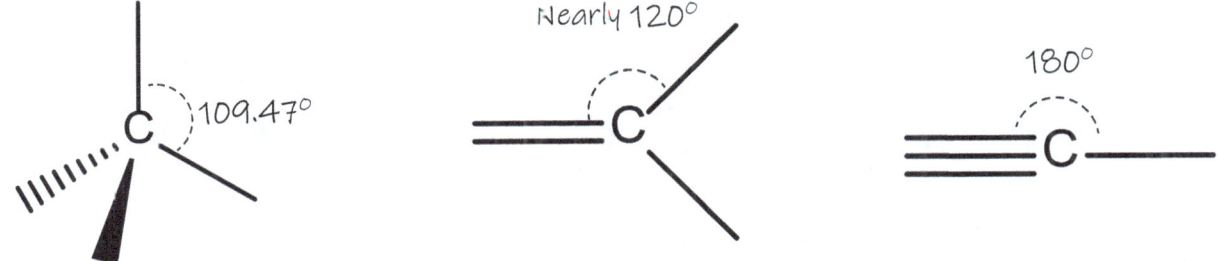

Several factors, such as hybridization of atomic orbitals, electron pairs repulsion, bond order, electronegativity of the bonded atoms, etc. are responsible for the deviation of bond angles. Thus in most of the cases, bond angles deviate from the ideal values, 109.47 degree (for single bonded carbon), or 120 degree (for double bonded carbon). However, at this moment, you are not ready to learn those complicated factors, which regulate the bond angles in organic molecules. But, you got the basic idea how bond angles vary when a multiple bond is associated in a carbon atom.

Hydrocarbons

The class of organic molecules called hydrocarbons consist of hydrogen and carbon atoms.
Methane is the simplest one because it contains one carbon atom only.

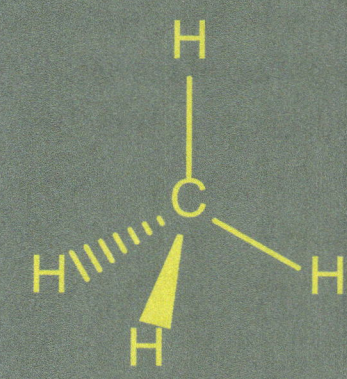

In the molecular structures of higher hydrocarbons, you find more than one carbon atom.

Ethane   Propane   Butane

How do we construct the structures of higher hydrocarbons?

Imagine a methane molecule lacking one hydrogen atom.

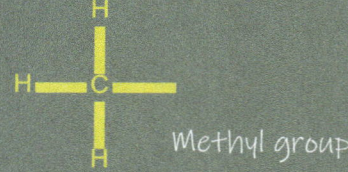

Methyl group

Notice that the above group of atoms is not a real molecule because the central carbon atom does not satisfy its four covalent bonds.

In organic chemistry, the shown group of atoms is known as the methyl group because it has one less hydrogen atom than the methane has.

You also imagine a methyl group as the previous way. The methyl group has one hand.

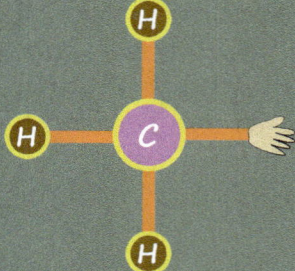

When two methyl groups make a carbon-carbon single bond, we get the structure of ethane.

Similarly, you may draw the structure of an ethyl group removing one hydrogen atom from ethane.

When a methyl group and an ethyl group make a carbon-carbon bond, we get the structure of propane.

In a similar way, you may draw the molecular structures of higher hydrocarbons containing greater numbers of carbon atoms.

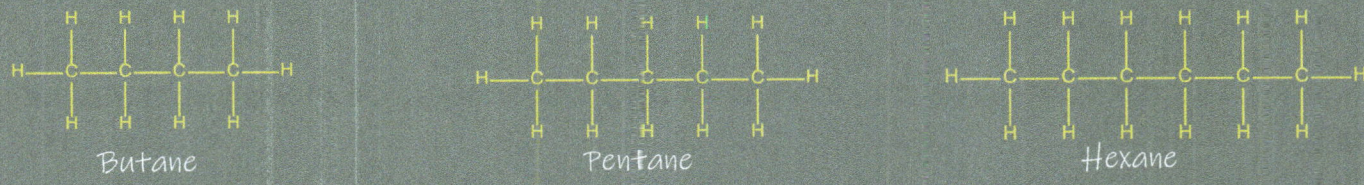

These hydrocarbons do not possess any double bond or triple bond. Such hydrocarbons are commonly called alkanes.

NOMENCLATURE OF ALKANES

We already discussed that millions of organic compounds may exist. If we called each of them by individual names, it would be difficult for us to remember those millions names. Thankfully, scientists devised a systematic technique to name organic compounds, so that one can easily determine the name of an organic compound from its molecular structure. This technique is commonly known as IUPAC nomenclature of organic compounds.

We mentioned the names of a few alkanes, such as methane, ethane, propane, butane, etc. Now we are going to reveal how they got such names.

These names are combination of two parts,

Methane → meth + ane
Ethane → eth + ane
Propane → prop + ane
Butane → but + ane
Pentane → pent + ane

The second part, 'ane' is common for all the cases. That part suggests that these organic molecules belong to alkane, which are hydrocarbons without carbon-carbon double bond or triple bond.

The first parts, meth, eth, prop, but, etc. denote the number of carbon atoms present in the chain.

When the number of carbon atom is 1, the prefix is meth.

When the number of carbon atoms is 2, the prefix is eth.

When the number of carbon atoms is 3, the prefix is prop.

When the number of carbon atoms is 4, the prefix is but.

When the number of carbon atoms is 5, the prefix is pent.

When the number of carbon atoms is 6, the prefix is hex.

When the number of carbon atoms is 7, the prefix is hept.

When the number of carbon atoms is 8, the prefix is oct.

When the number of carbon atoms is 9, the prefix is non.

When the number of carbon atoms is 10, the prefix is dec.

Now it is easy to determine the name of any alkane, which consists of a long carbon chain made of a given number of carbon atoms.

For example, the name of an alkane, which is eight carbon atoms long, will be octane.

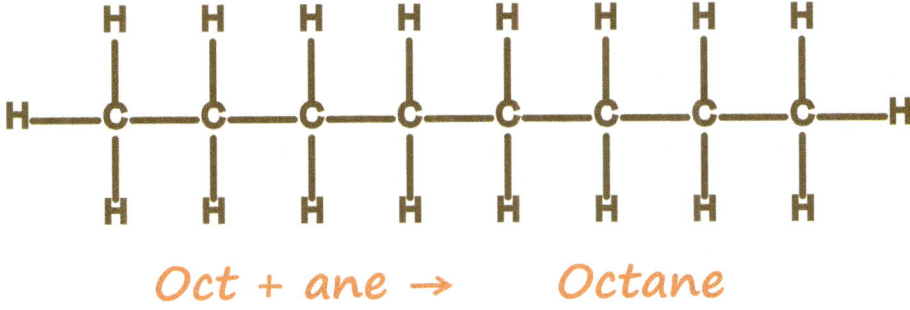

Oct + ane → Octane

But many alkane molecules do not always have linear carbon chains. They may have branched carbon skeletons.

The following alkanes have branched carbon skeletons.

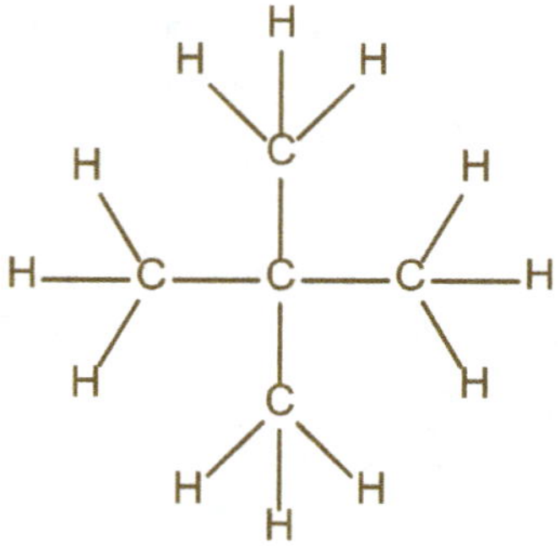

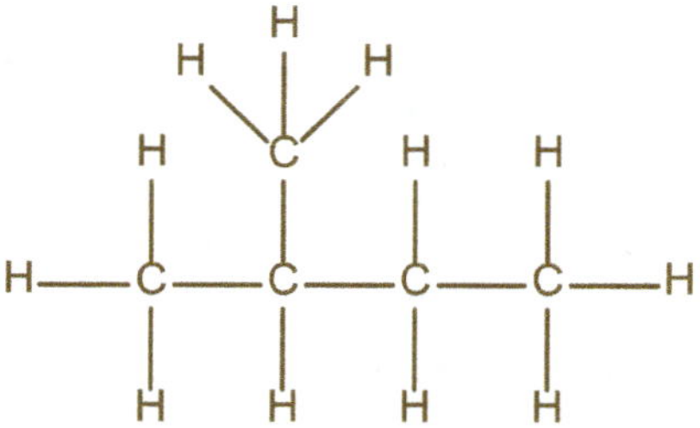

In order to name those compounds, some additional rules are required. Here we are going to discuss those rules citing a few examples.

Consider the following branched alkane molecule.

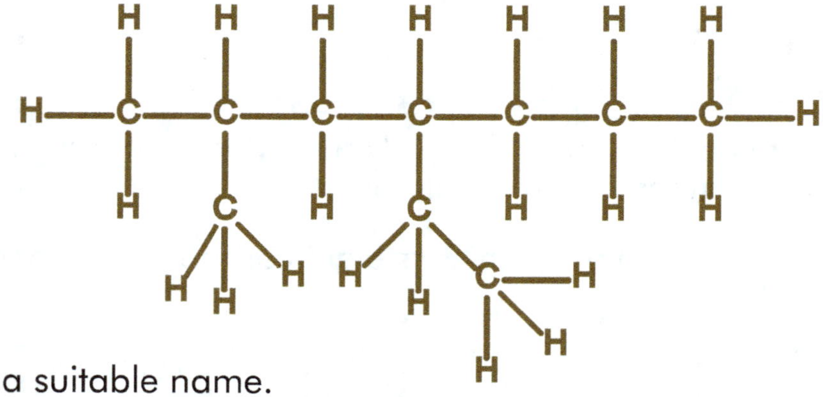

We have to give it a suitable name.

In order to do that, first we need to find the longest carbon chain.

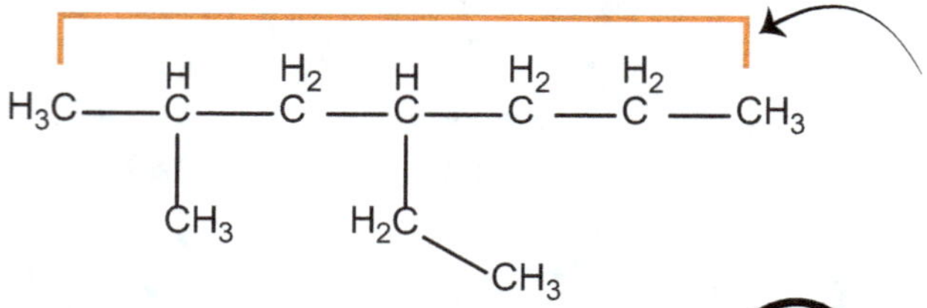

The longest carbon chain is seven carbon atoms long.

The next step is to number the carbon atoms 1 to 7 (because the chain is seven carbon atoms long). We have to keep in mind that the numbering should be in the direction, in which the substituted carbon atom falls at the smallest possible number.

This is the correct direction of numbering because here the substituted carbon atom comes at the number 2.

If we put the numbers in the other direction, it would be wrong.

In that case, the substituted carbon atom would come at number 4, which is not the smallest possible number.

Now notice, there is a methyl group bonded to the carbon number 2, and an ethyl group at the carbon number 4. Also, the longest carbon chain is 7 carbon atoms long, which corresponds to heptane.

Therefore, the IUPAC name is,

4-ethyl-2-methylheptane.

(Notice, ethyl and methyl are written in the alphabetical order.)

There are several other rules that have to be considered while writing the names of larger organic compounds. However, we are not going to discuss those rules further, otherwise you will lose interest in organic chemistry very soon.

How to draw the structures of large organic molecules

We discussed that four covalent single bonds around a carbon atom stay in a tetrahedral arrangement in space. The three dimensional picture of four C-H bonds in methane can be shown by wedge-dash notation.

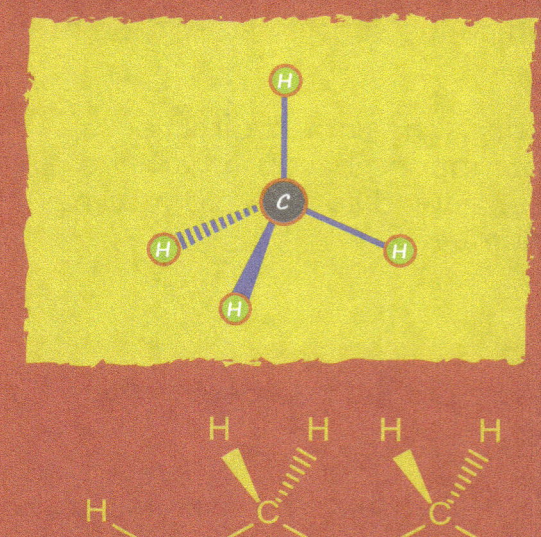

In a similar way, it is possible to draw the three dimensional structures of higher alkanes.

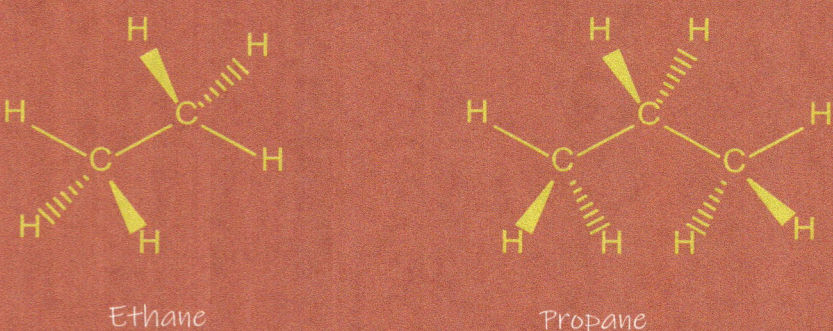

Ethane   Propane   Butane

However, it is quite a cumbersome job to show all the carbon and hydrogen atoms along with C-H bonds while drawing the structure of a large organic molecule. For that reason, organic molecules are often shown by writing skeletal formulas where the C-H bonds and atomic symbols of carbon and hydrogen are not shown.

The structure of propane molecule can be shown as,

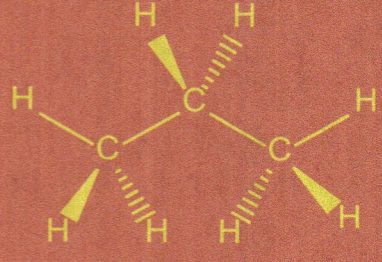 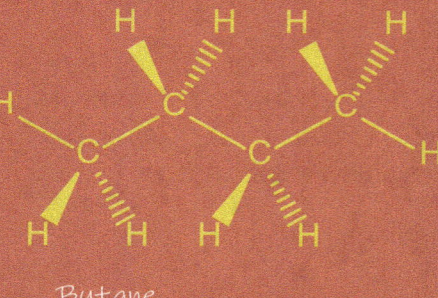

The end points and the junctions of the straight lines represent the carbon atoms that are covalently bonded with appropriate numbers of hydrogen atoms.

Any organic molecule can be represented by its skeletal formula. Here the skeletal formulas of some alkanes are shown.

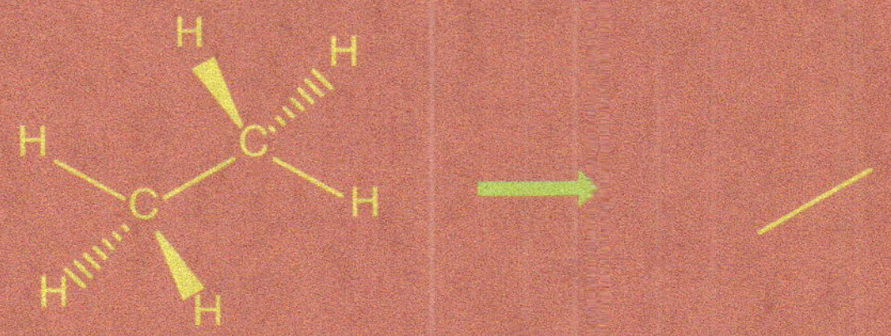

Ethane

Propane

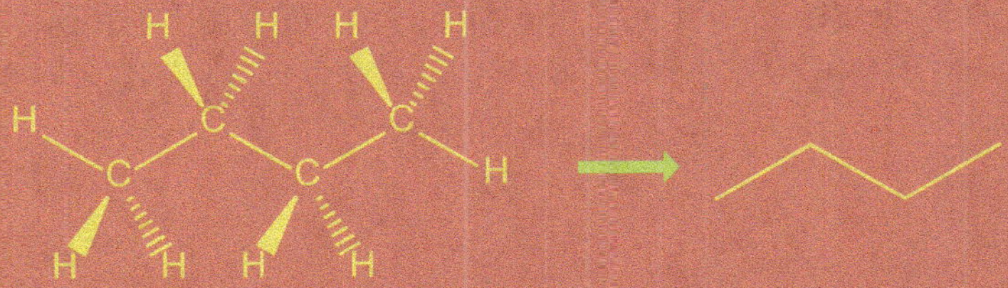

Butane

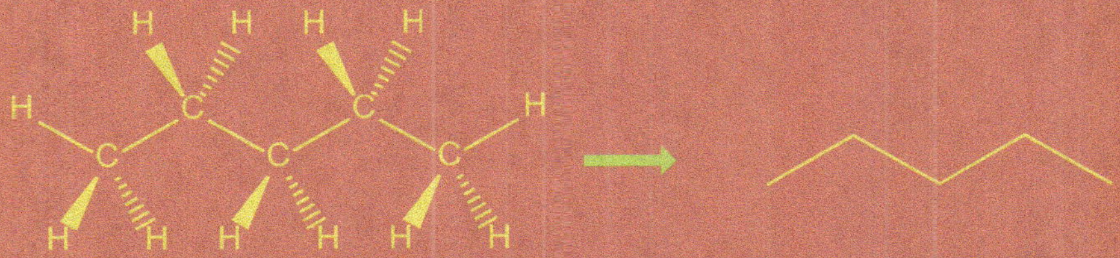

Pentane

Other Hydrocarbons

Alkanes do not possess any double bond or triple bond in their molecular structures. However there are many hydrocarbons that contain carbon-carbon double bonds and triple bonds.

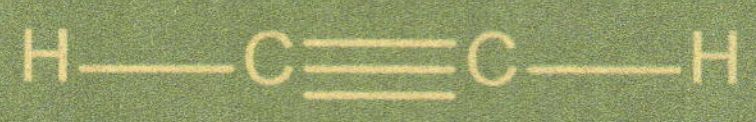

Acetylene

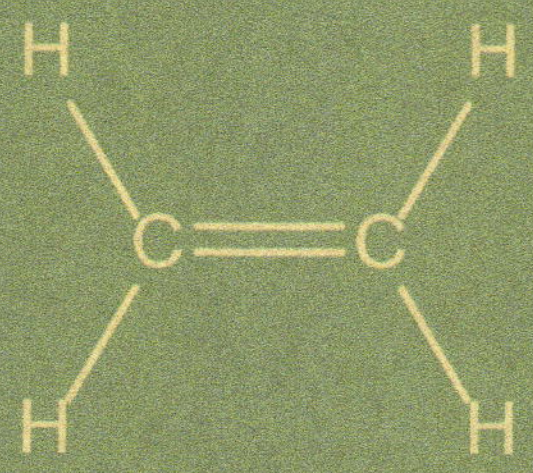

Ethylene

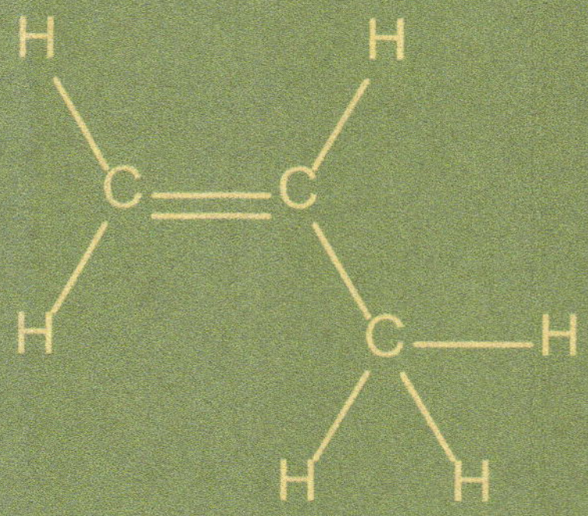

Propylene

The double bond and triple bond containing hydrocarbons are often called unsaturated hydrocarbons, because the double bonds and the triple bonds are available for further bond formation. When they react with hydrogen, alkanes are formed.

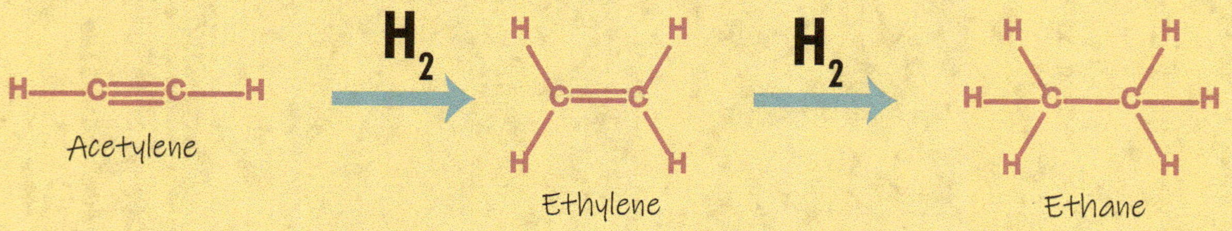

Acetylene $\xrightarrow{H_2}$ Ethylene $\xrightarrow{H_2}$ Ethane

On the other hand, alkanes are called saturated hydrocarbons because they have all single bonds, which are quite unreactive.

You know that the names of alkanes end with the suffix 'ane'. Likewise, carbon-carbon double bonded and carbon-carbon triple bonded hydrocarbons end their names with suffixes 'ene' and 'yne' respectively.

Carbon-carbon double bonded hydrocarbons are called alkenes, and carbon-carbon triple bonded hydrocarbons are called alkynes.

Now we are able to determine the names of unsaturated hydrocarbons.

This hydrocarbon has two carbon atoms, so the prefix becomes 'eth', and it has a double bond, so the suffix becomes 'ene'. There the IUPAC name of this hydrocarbon is ethene.
However the common name of this hydrocarbon is ethylene.

eth + ene → ethene

This hydrocarbon has two carbon atoms, so the prefix becomes 'eth', and it has a triple bond, so the suffix becomes 'yne'. There the IUPAC name of this hydrocarbon is ethyne.
However the common name of this hydrocarbon is acetylene.

eth + yne → ethyne

prop + ene → propene

prop + yne → propyne

When the structure of an alkene or alkyne contains more than three carbon atoms, you need to specify the position of the double bond or triple bond, because more than one molecular structure is possible in such cases.

Both of these structures represent the isomers of butene. They contain equal numbers of carbon and hydrogen atoms, but the positions of the double bonds are different.

The numbering on the carbon atoms can specify the positions of the double bonds.

What are isomers?

Isomers are two or more chemical compounds possessing identical chemical formulas, but different molecular structures.

1-butene

2-butene

1-butene and 2-butene have identical chemical formulas, C_4H_8, but they have different molecular structures. Therefore, they are isomers of each other.

CIS TRANS ISOMERISM

2-butene can show another pair of isomerism called cis- trans- isomerism.

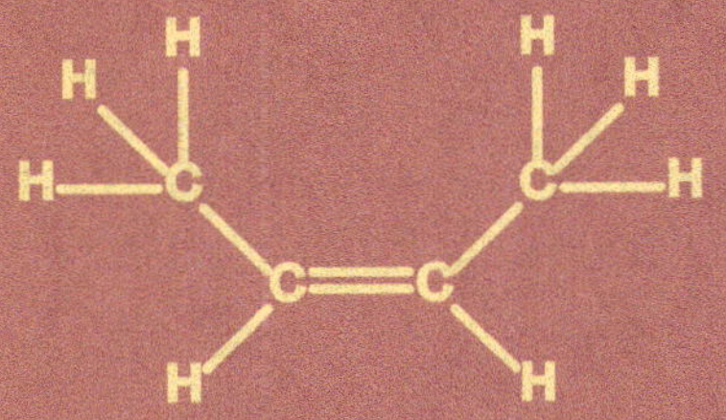

cis 2-butene

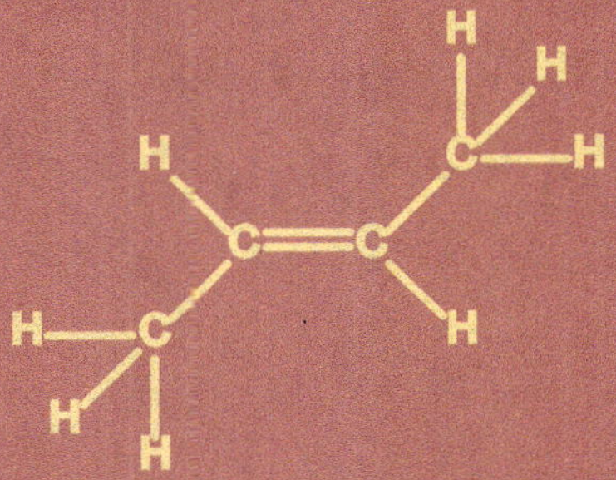

trans 2-butene

In the cis isomer, the hydrogen atoms and the methyl groups belong to the same sides of the double bond, whereas in the trans isomer, the hydrogen atoms and the methyl groups belong to the opposite sides.

Some unsaturated hydrocarbons may possess more than one double bond or triple bond.

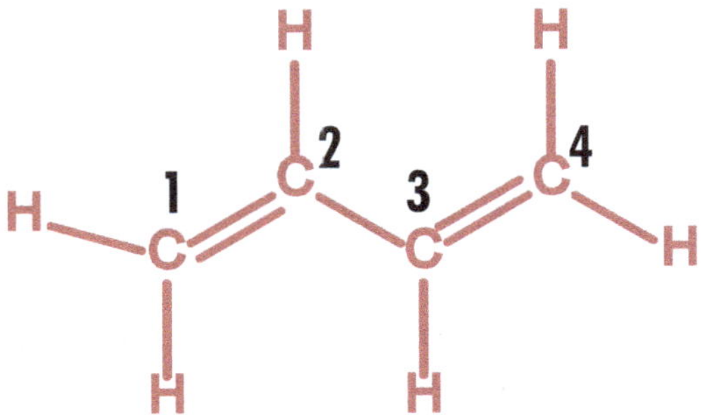

1,3-butadiene

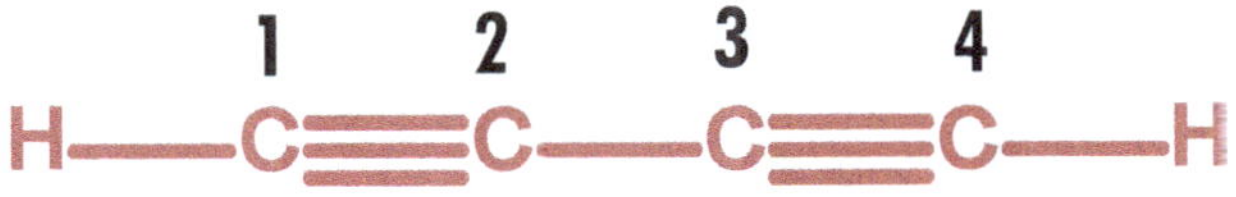

1,3-butadiyne

In such cases, the number of double bonds or triple bonds are specified by diene, triene, or diyne, triyne etc.

Cyclic hydrocarbons

Cyclic hydrocarbons are also common in organic chemistry. Let's see how we draw the structures of cyclic hydrocarbons ourselves.

Consider the structure of a propane molecule.

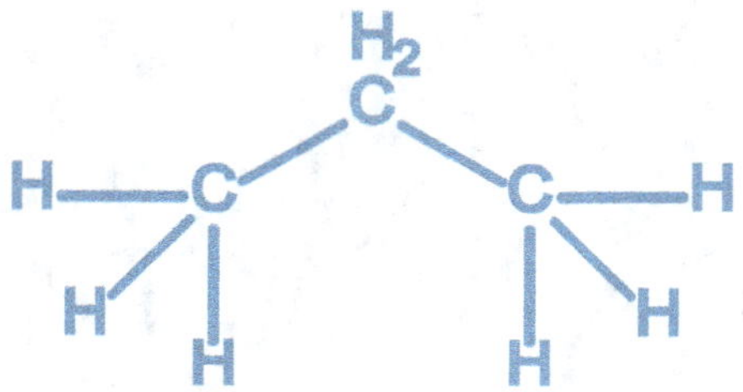

If we remove two hydrogen atoms from the terminal carbons, we get the following structure.

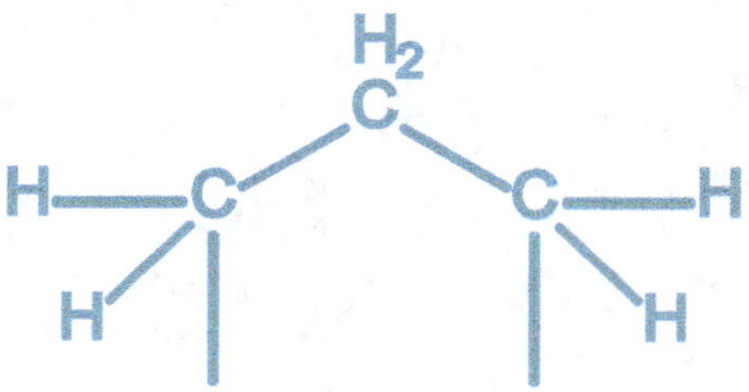

Notice the terminal carbon atoms do not satisfy four single covalent bonds. If these two carbon atoms make a single bond between themselves, all the carbon atoms will satisfy their four covalent bonds.

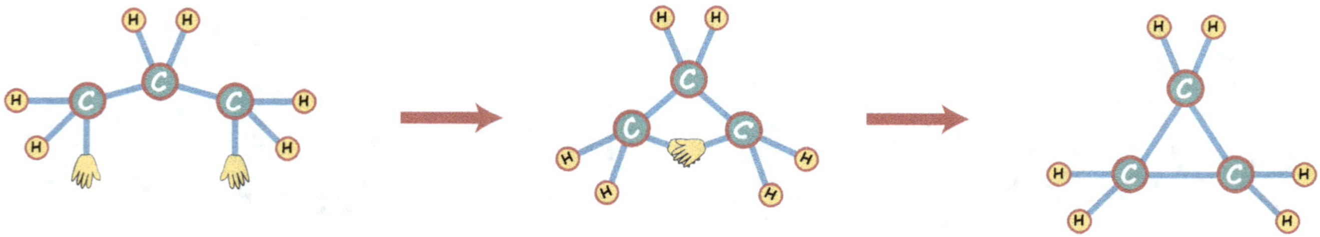

Therefore, we get a cyclic molecule called cyclopropane.

The prefix, 'cyclo' is added before propane in order to specify that the said alkane is a cyclic alkane.

Other cyclic alkanes are also known.

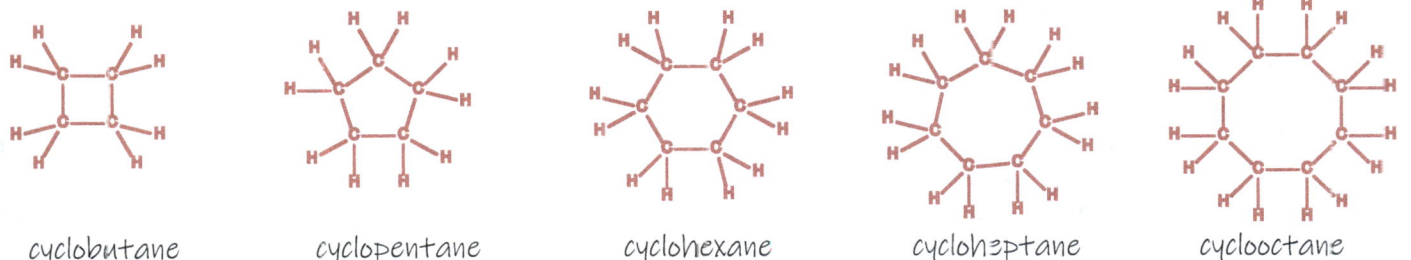

cyclobutane   cyclopentane   cyclohexane   cycloheptane   cyclooctane

These cycloalkanes are generally represented by their skeletal formulas.

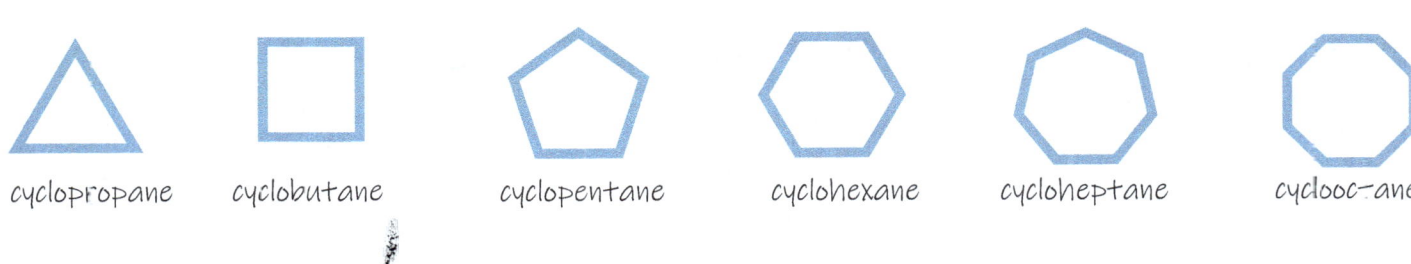

cyclopropane   cyclobutane   cyclopentane   cyclohexane   cycloheptane   cyclooctane

When cyclic organic compounds possess carbon-carbon double bonds, we call them cycloalkenes.

cyclopropene   cyclobutene 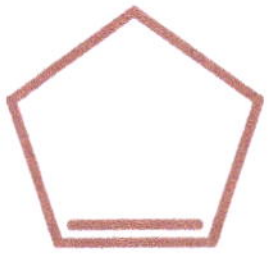   cyclopentene 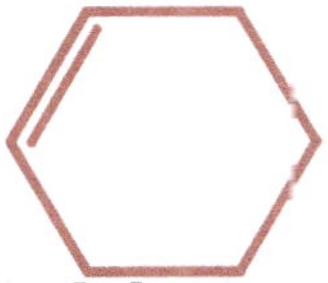   cyclohexene

cyclobutadiene 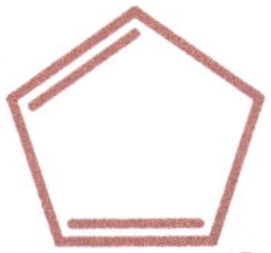   cyclopentadiene 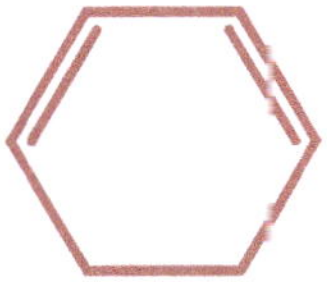   1,3-cyclohexadiene 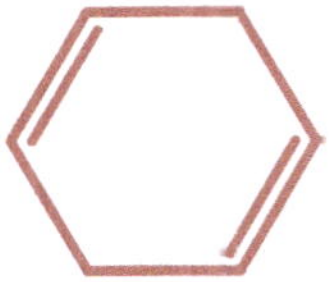   1,4-cyclohexadiene

Benzene

The cyclic hydrocarbon possessing six carbon atoms and three alternate double bonds is quite unique.

The compound is called benzene.

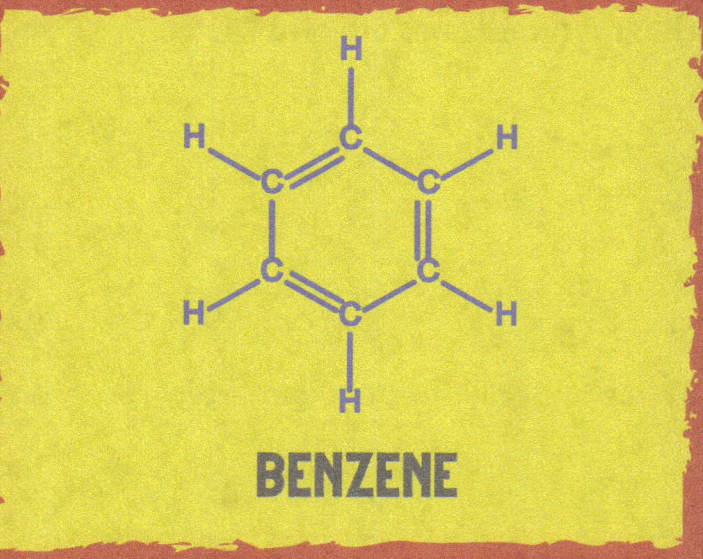

BENZENE

The three double bonds do not stay fixed between the specific pairs of carbon atoms, rather they can alter their positions to the alternate carbon pairs.

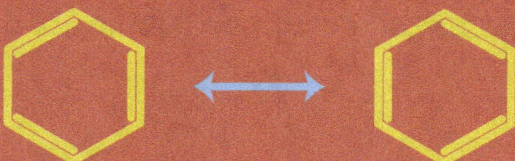

Let's discuss why it is possible.

We know that every covalent bond is the result of sharing a pair of electrons. Therefore, a carbon-carbon double bond is the result of sharing two pairs of electrons.

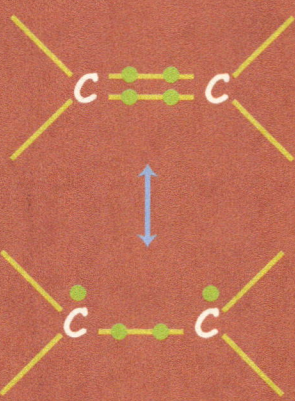

When one of these two bonds cleaves homolytically, there will be one lone electron on each of the carbon atoms.

The same phenomenon happens in the benzene molecule.

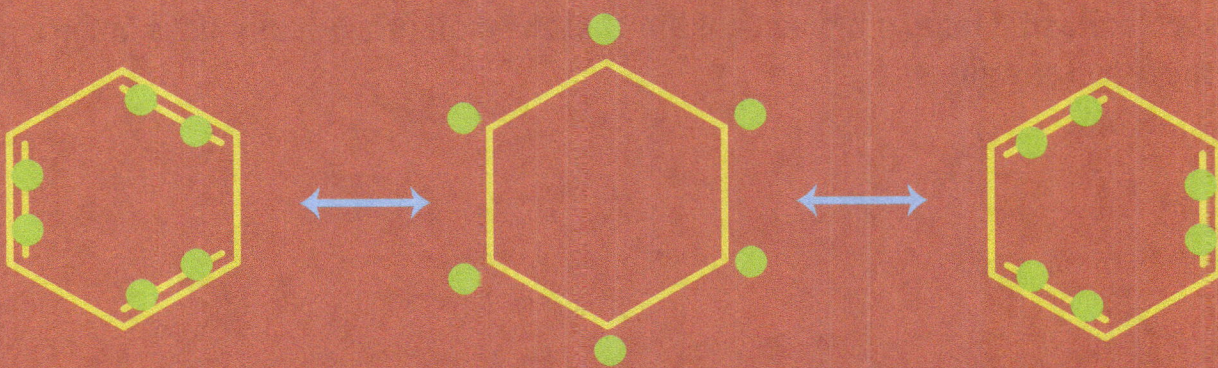

For that reason, the carbon-carbon double bonds along with the associated electrons are not localized in the benzene molecule, and both the structures are equally possible. Such equivalent structures are called the canonical structures of benzene.

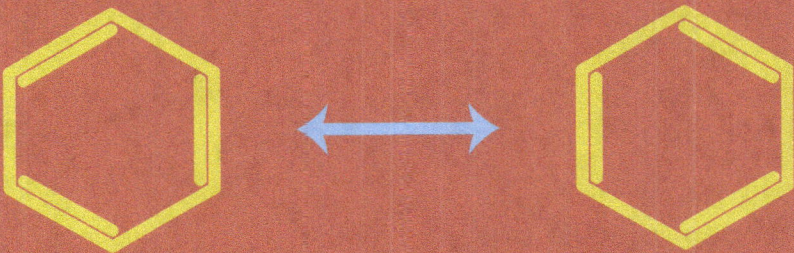

This phenomenon is known as resonance. Several organic molecules that have alternate single bond-double bond systems show such phenomenon.

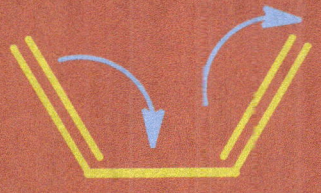

Resonance increases the stability of a molecule because the electrons get more space to move.
According to quantum physics, when a particle gets more space to delocalize itself, the energy of the system becomes lower, and the lowering of energy means gain of extra stability.

Benzene has two canonical structures. It does not mean that the benzene molecule shows one of these two structures in a certain moment, but both the structures are equally possible for a benzene molecule at a time. It means that the benzene molecule actually stays in the average form of these two structures.

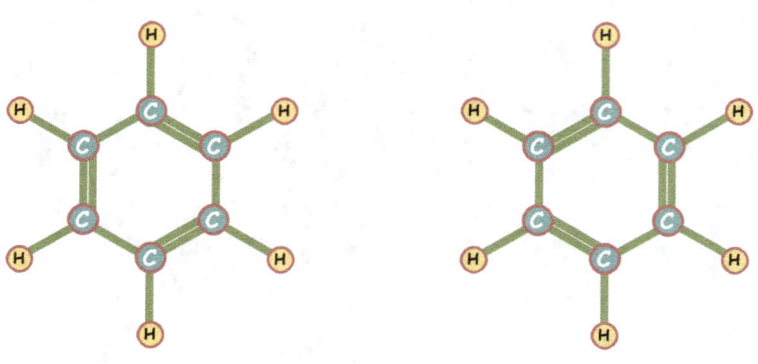

For that reason, each C-C bond in the benzene molecule is neither a true single bond, nor a true double bond. We may say that every C-C bond in the benzene molecule is a 1.5 bond (one and half bonds).

Moreover, due to the average structure, all the C-C bonds in a benzene molecule are identical, and the benzene molecule is often represented by the following structure.

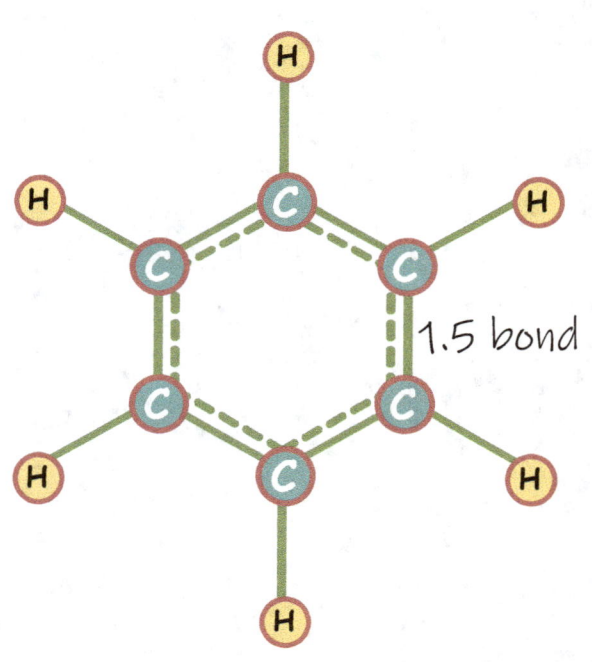

1.5 bond

The benzene molecule is an aromatic compound because it shows aromaticity. Keep in mind that the aromaticity is not related to aroma, this is a special property of planar cyclic organic compounds that show cyclic resonance through delocalization of $4n+2$ electrons (2, 6, 10, etc.) like benzene (benzene is associated with delocalization of 6 electrons).

The benzene ring also known as phenyl ring is very common in various organic compounds.

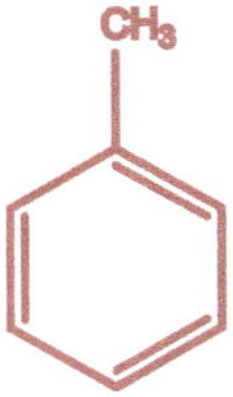

Toluene

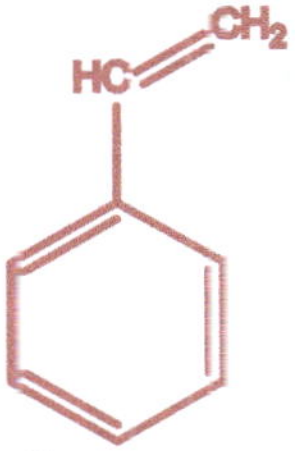

Styrene

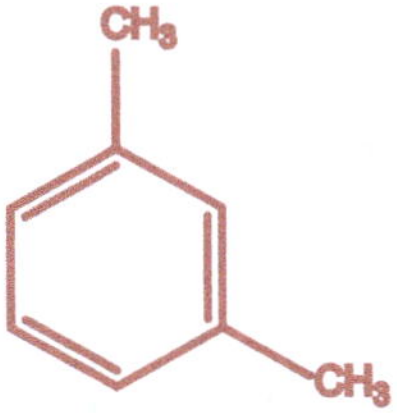

Meta xylene

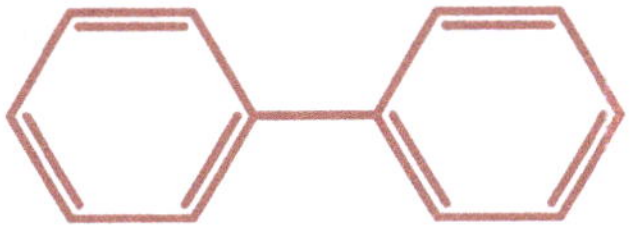

Biphenyl

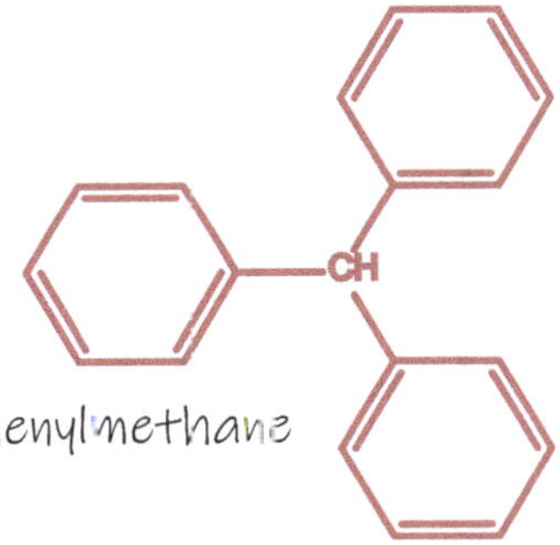

Triphenylmethane

The structure of an organic compound called naphthalene, which is commonly known as mothballs, consists of two fused benzene rings.

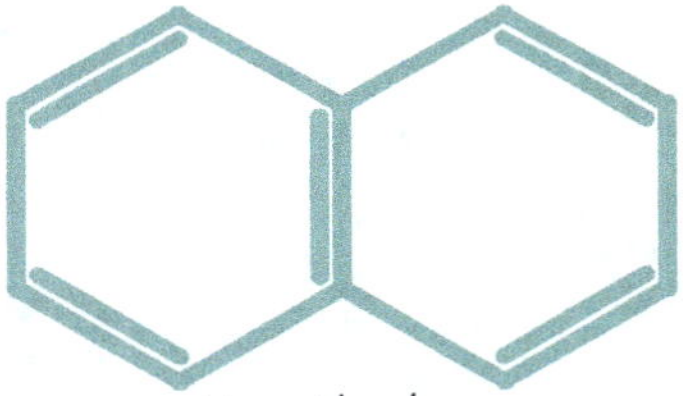

Napthalene

The fused benzene rings are also found in the structures of anthracene and phenanthrene.

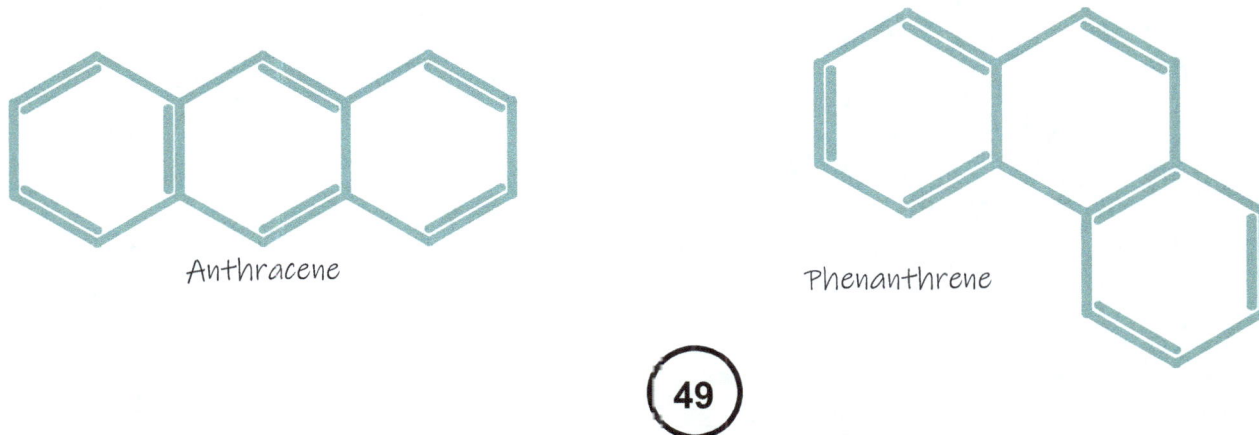

Anthracene

Phenanthrene

FUNCTIONAL GROUPS

So far we have discussed hydrocarbons and their structures. You know that the molecules of hydrocarbons consist of carbon and hydrogen atoms only. But many organic compounds may contain other elements, such as oxygen, nitrogen, sulfur, halogens (fluorine, chlorine, bromine, iodine), etc. along with carbon and hydrogen.

One or more atoms of these elements form functional groups, and the presence of these functional groups in an organic molecule determines the chemical and other properties of that compound.

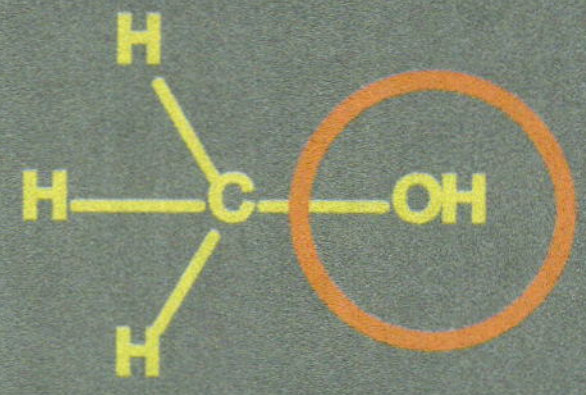

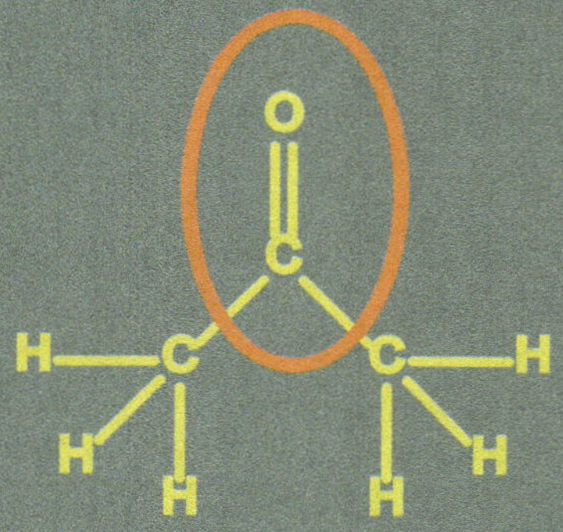

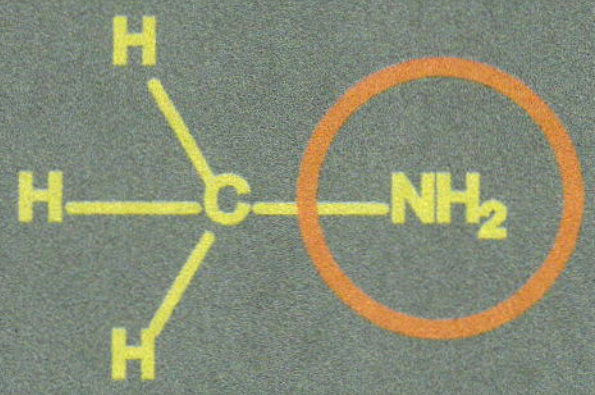

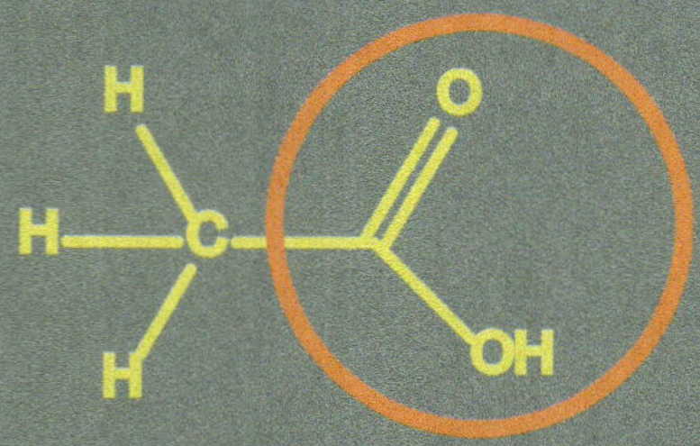

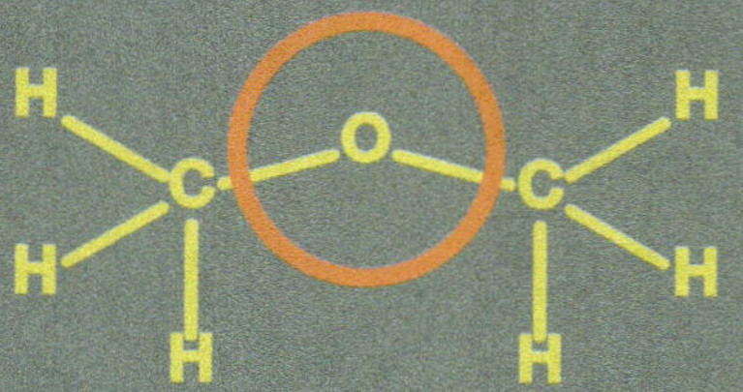

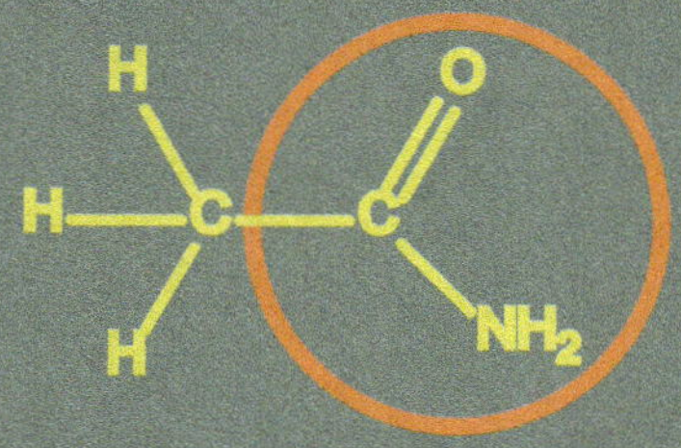

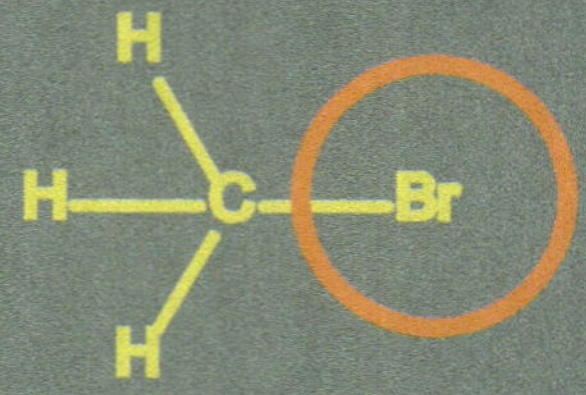

FUNCTIONAL GROUP FOR ALCOHOLS

The Hydroxyl Group (-OH)

You should remember the fact that an oxygen atom is able to form two covalent bonds.

Likewise, a hydrogen atom is able to form one covalent bond.

If an oxygen atom and a hydrogen atom make a covalent bond between them, you get the following structure.

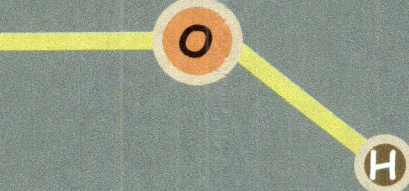

The above structure represents the hydroxyl functional group. Notice, among the two sites of the oxygen atom, which are available for covalent bonding, one site is satisfied by a hydrogen atom making a O-H covalent bond, and other site is still available for further bonding.

Like a hydroxyl group, an alkyl group also possesses a vacant site for covalent bonding.

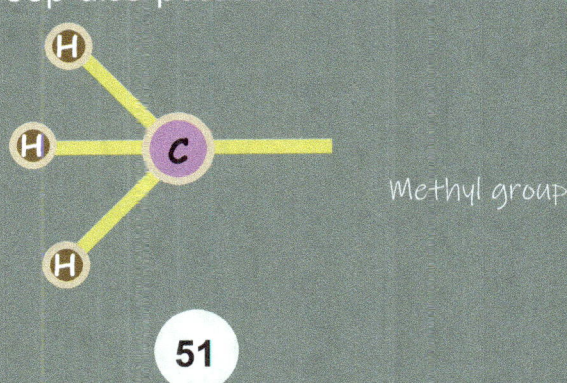

Methyl group

When the vacant site of an alkyl group is satisfied by a hydroxyl functional group, an alcohol is formed.

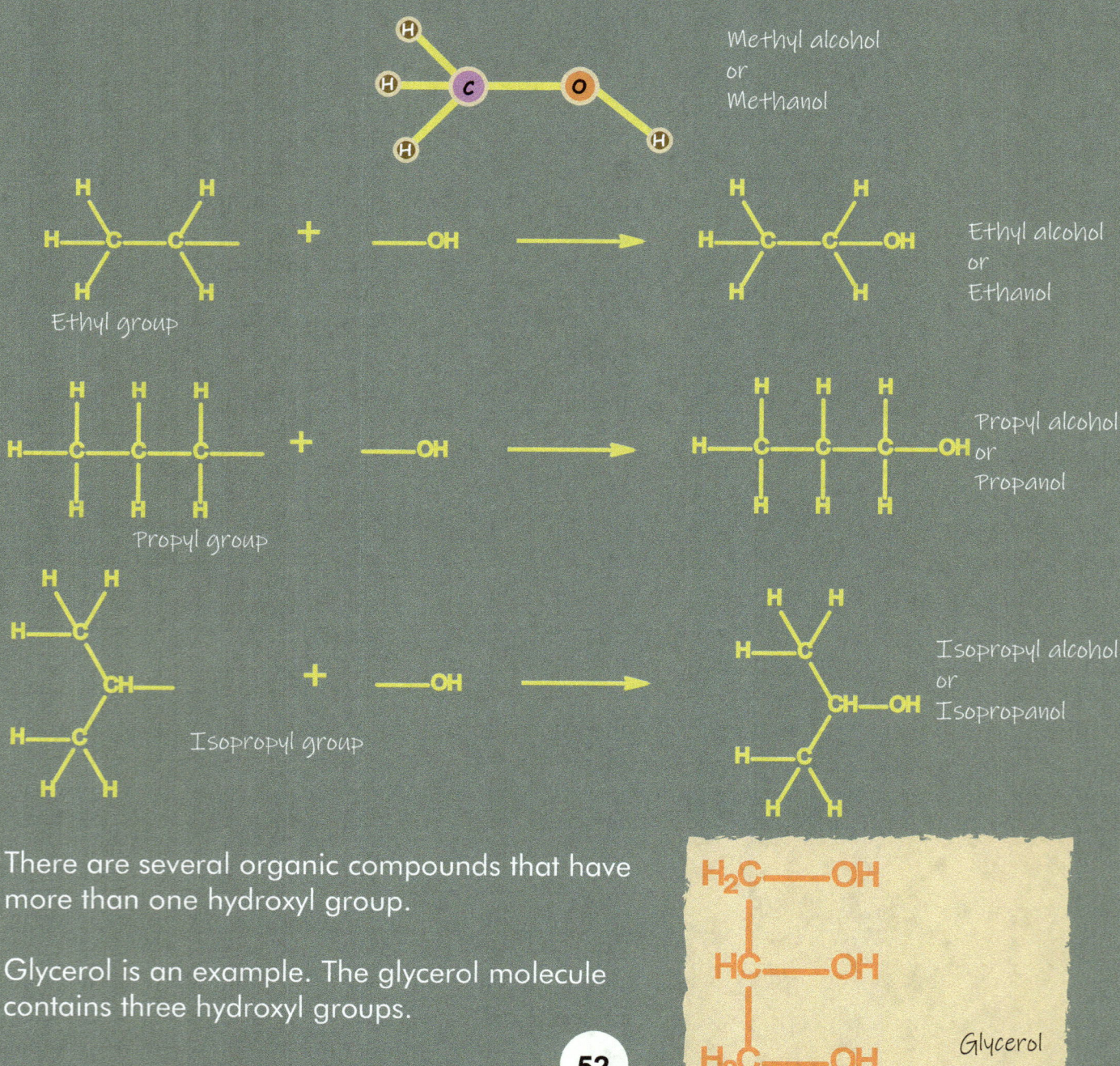

There are several organic compounds that have more than one hydroxyl group.

Glycerol is an example. The glycerol molecule contains three hydroxyl groups.

FUNCTIONAL GROUP FOR ALDEHYDES AND KETONES

The Carbonyl Group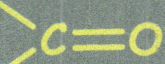

You know that the carbon atom makes 4 covalent bonds, and the oxygen atom makes 2 covalent bonds with other elements. This concept can be represented as,

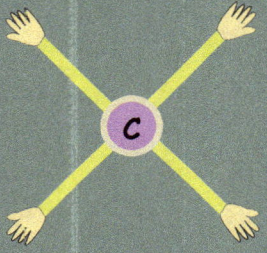
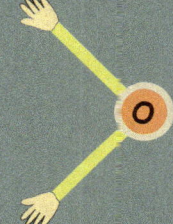

The carbon atom has 4 hands.   The oxygen atom has 2 hands.

The two available hands of the oxygen atom may hold two hands of carbon among the four in the following way,

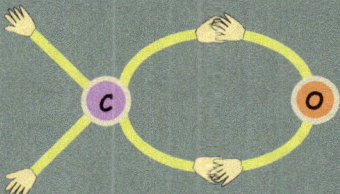

It actually represents one carbon-oxygen double bond along with two available sites on the carbon atom for further bonding.

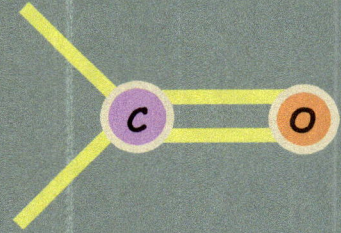

The functional group is known as the carbonyl group.

53

The two available sites on the carbon atom of a carbonyl group may form bonds with hydrogen atoms or alkyl groups.

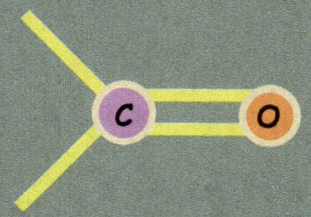

When one of the two sites is satisfied by a hydrogen atom, and the other is satisfied by an alkyl group (or another hydrogen atom), the compound is called an aldehyde.

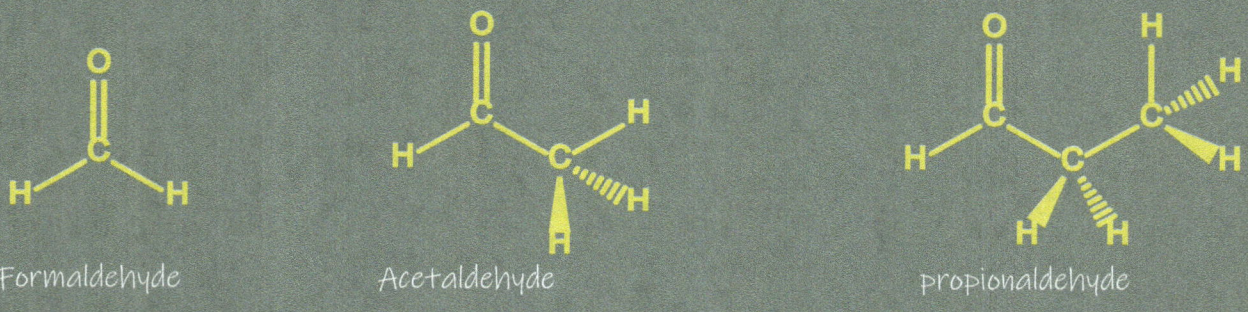

Formaldehyde Acetaldehyde Propionaldehyde

On the other hand, when both the bonds are formed with alkyl groups, the compound is called a ketone.

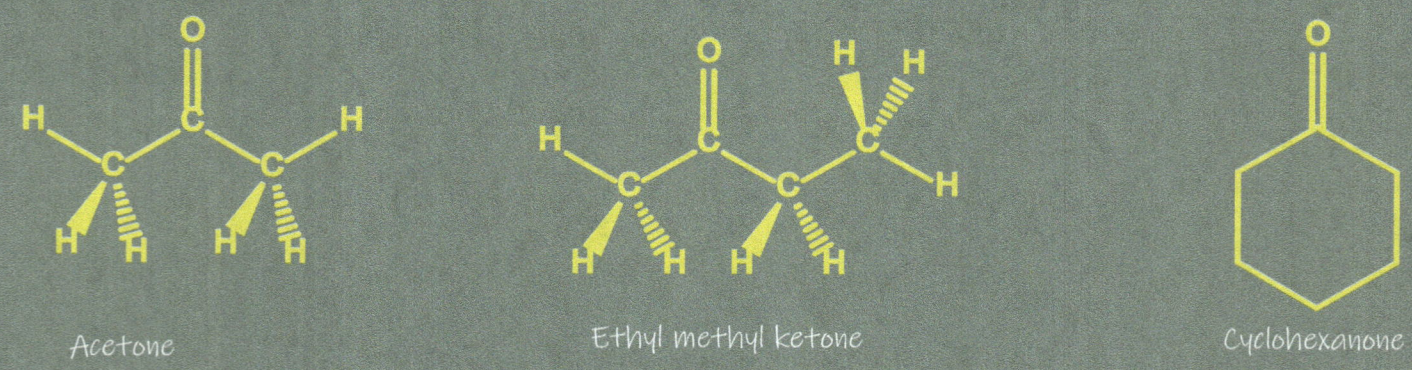

Acetone Ethyl methyl ketone Cyclohexanone

FUNCTIONAL GROUP FOR CARBOXYLIC ACIDS

The Carboxyl Group

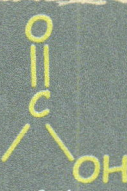

When one of the available sites of the carbonyl group makes covalent bond with a hydroxyl group, we get a carboxyl functional group.

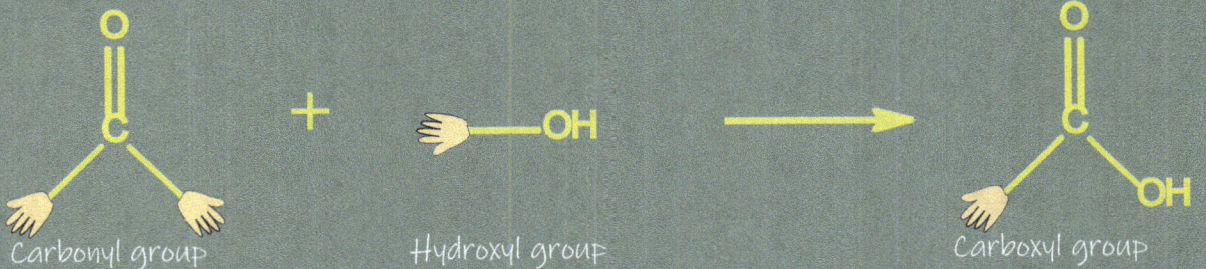

The carboxyl group has one available site to make a covalent bond. The available site may form a covalent bond with a hydrogen atom or an alkyl group.

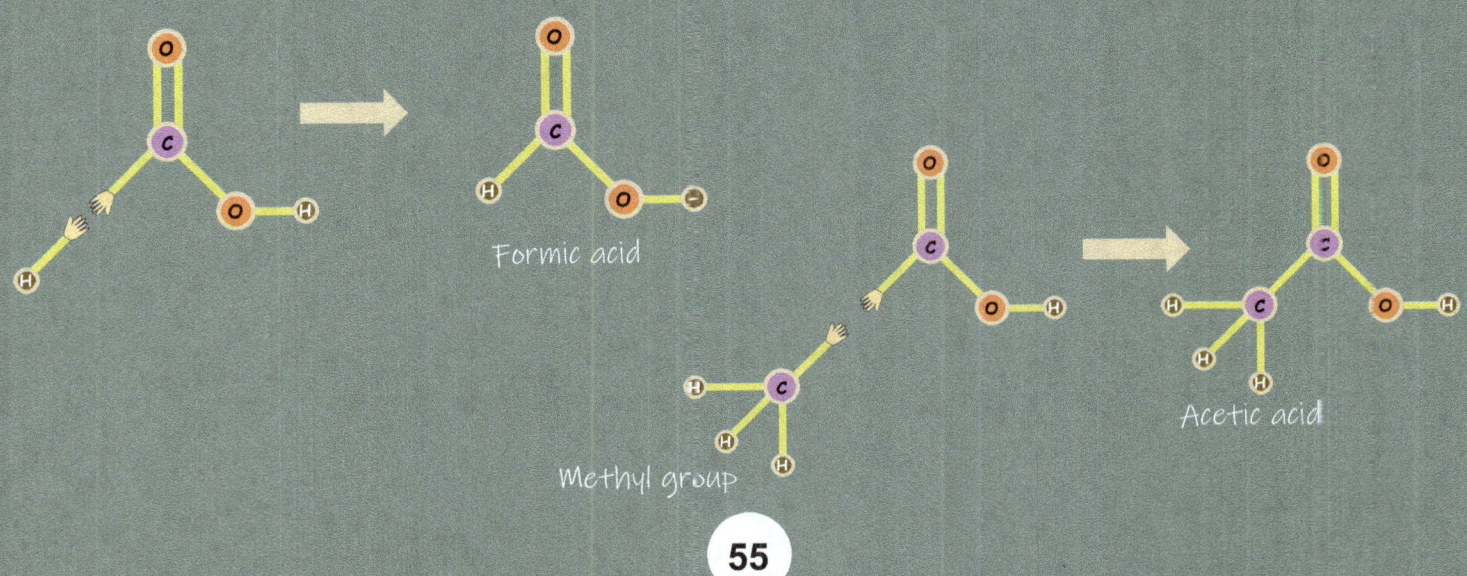

The organic compounds containing more than one carboxyl group are also known.

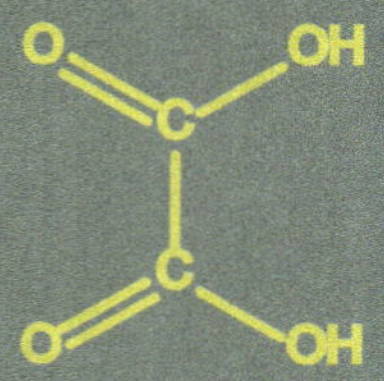

Oxalic acid

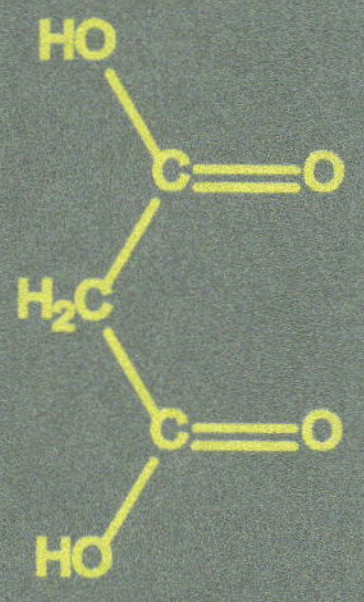

Malonic acid

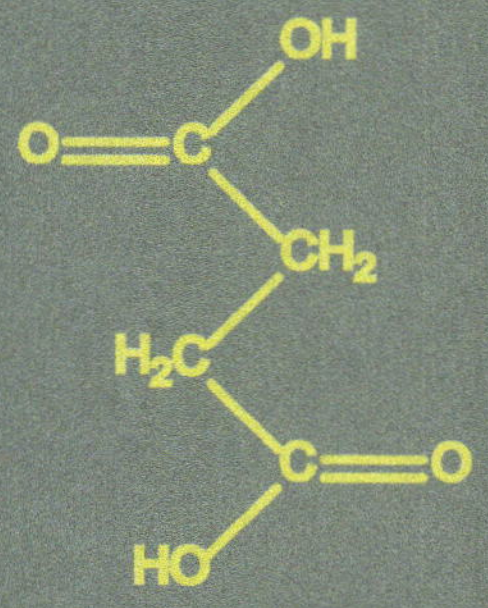

Succinic acid

The organic compounds containing one or more carboxyl groups are termed as carboxylic acids. The compounds are classified acids because the carboxyl group releases a positively charged hydrogen ion in solution.

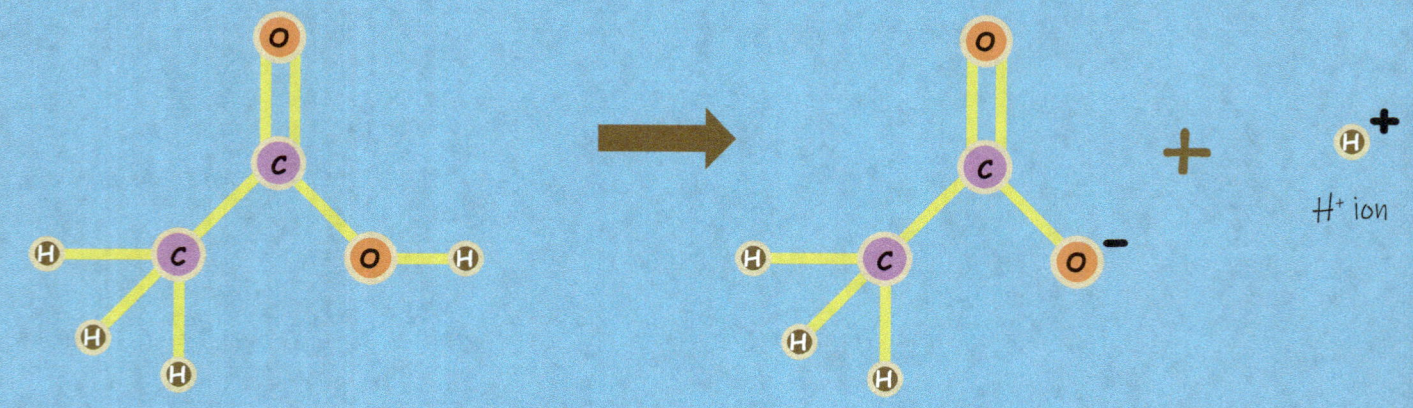

According to the definition, a hydrogen containing compound that releases H^+ ions in aqueous solution is called an acid. Carboxylic acids also do the same.

FUNCTIONAL GROUP FOR ETHERS

The Ether Group

The functional group of ether molecules is simple. It is just an oxygen atom with its two available sites for covalent bonding.

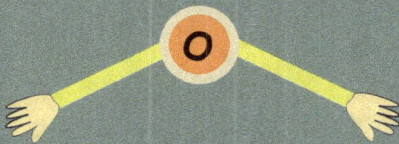

When two alkyl groups make covalent bonds with the oxygen, the resulting compound is called an ether.

Therefore, general molecular structure of ether compound is,

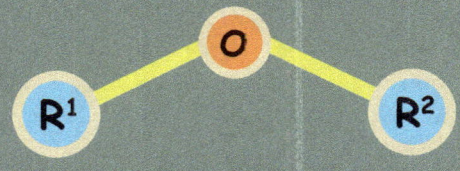

Where R^1 and R^2 represent two identical or different alkyl groups.

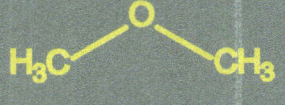

Dimethyl ether   Ethyl methyl ether   Diethyl ether

FUNCTIONAL GROUP FOR ESTERS

The Ester Group

When the hydrogen atom of a carboxylic acid group is replaced by an alkyl group, you get the structure of the ester group.

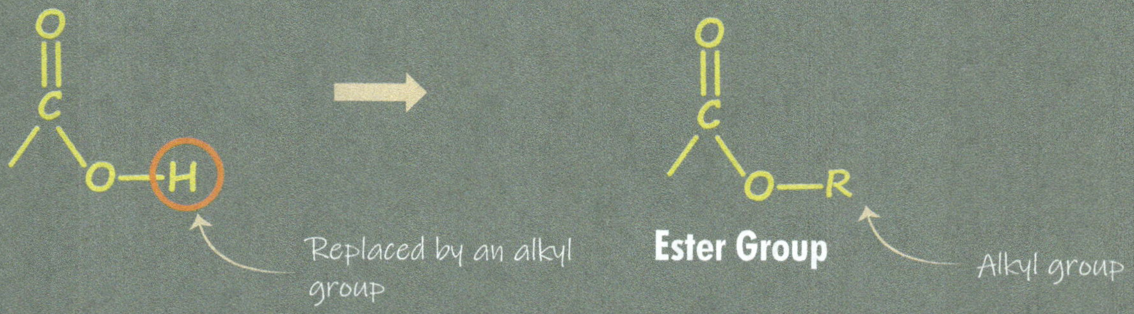

Replaced by an alkyl group → **Ester Group** (Alkyl group)

Therefore, an ester compound has the general formula,

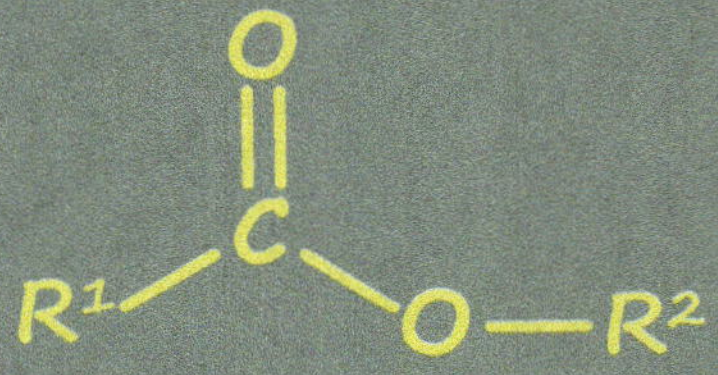

The examples of ester are,

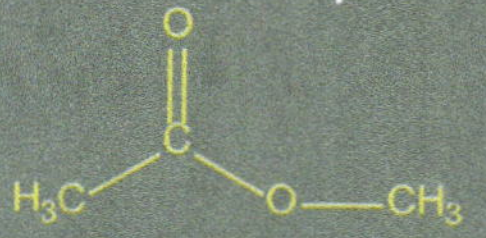

- Methyl acetate ($H_3C-COO-CH_3$)
- Ethyl acetate ($H_3C-COO-C_2H_5$)
- Methyl propanoate ($C_2H_5-COO-CH_3$)

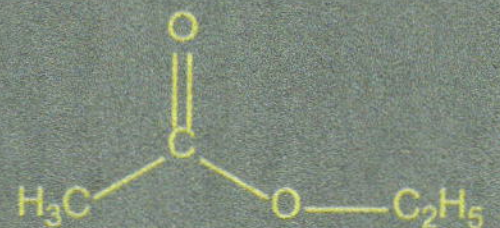

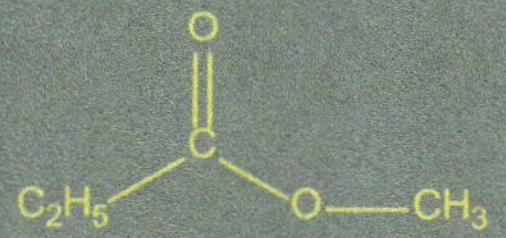

Where R^1 and R^2 are alkyl groups.

An ester compound is generally formed when a carboxylic acid reacts with an alcohol.

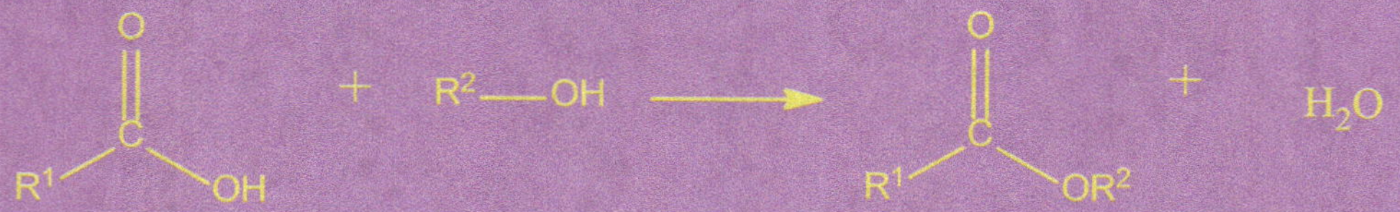

$$R^1COOH + R^2-OH \rightarrow R^1COOR^2 + H_2O$$

In chemistry, an acid and a base react with each other and produce salt and water. Here the carboxylic acid is an acid and the alcohol acts as a base, and when they react, an ester compound is produced along with water. For that reason, you may consider the ester compound as the organic salt, and the above reaction is known as esterification reaction.

FUNCTIONAL GROUP FOR AMINES

The Amine Group

The previously discussed functional groups were associated with oxygen atoms. Now it is time to discuss the functional groups containing nitrogen atoms.

The nitrogen atom is susceptible to make three covalent bonds with other atoms.

Therefore, up to 3 alkyl groups may join with a nitrogen atom through covalent bonding.

When a nitrogen atom makes a bond with one alkyl group, and the remaining two available sites are satisfied with two hydrogen atoms, the organic compound is called a primary amine (or 1 degree amine).

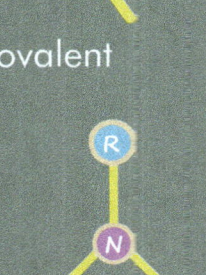
Primary amine

When a nitrogen atom makes bonds with two alkyl groups, and the remaining available site is satisfied with one hydrogen atom, the organic compound is called a secondary amine (or 2 degree amine).

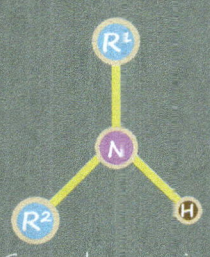
Secondary amine

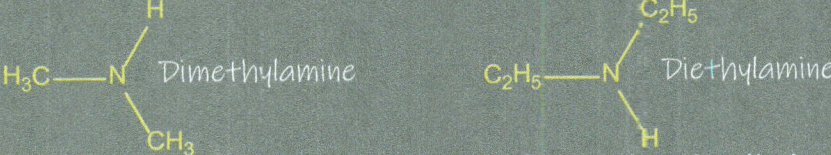

And when all the three bonds are formed with alkyl groups, the compound is known as a tertiary amine (or 3 degree amine).

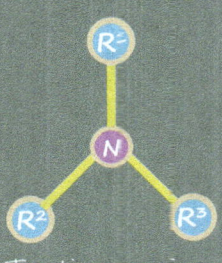
Tertiary amine

FUNCTIONAL GROUP FOR AMIDES

The Amide Group

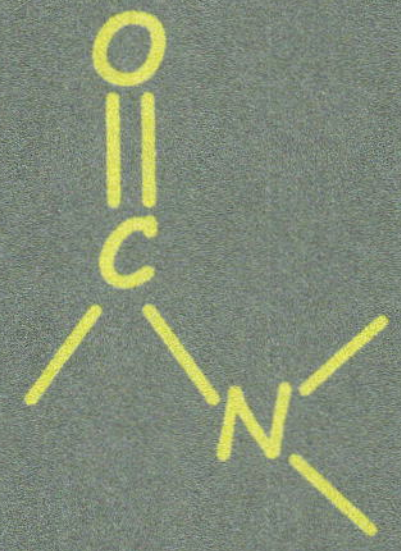

Amines are basic compounds because they produce OH- ions in aqueous medium. When an amine compound is treated with a carboxylic acid, a compound called amide is produced along with water.

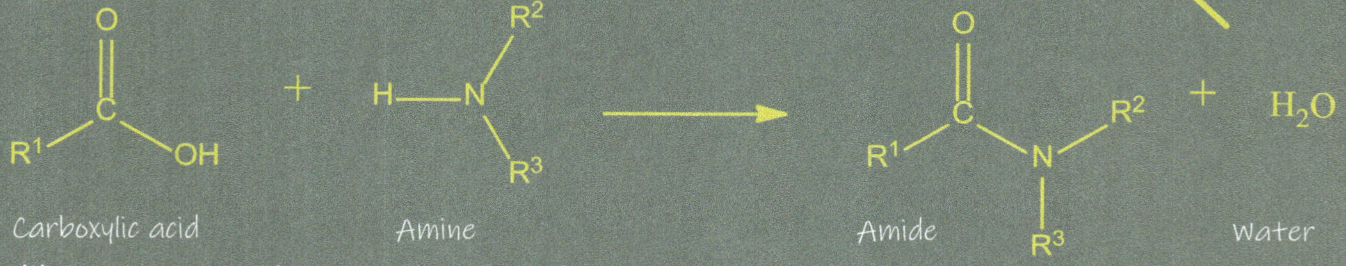

Carboxylic acid + Amine → Amide + H_2O

Just like esters, amides are also classified as organic salts.

The general structures of amides are:

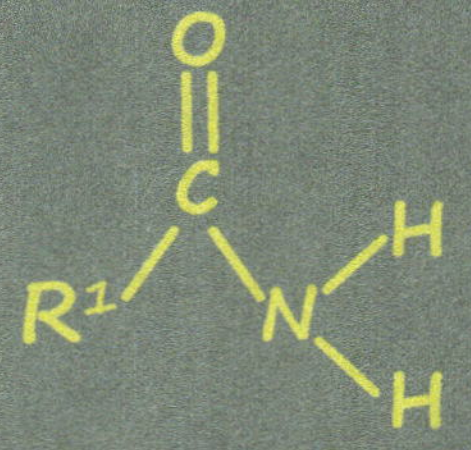

Primary amide

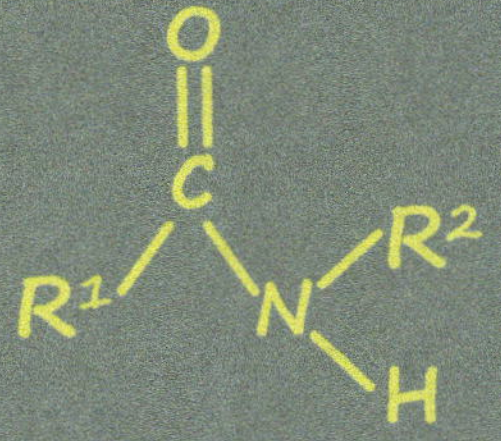

Secondary amide

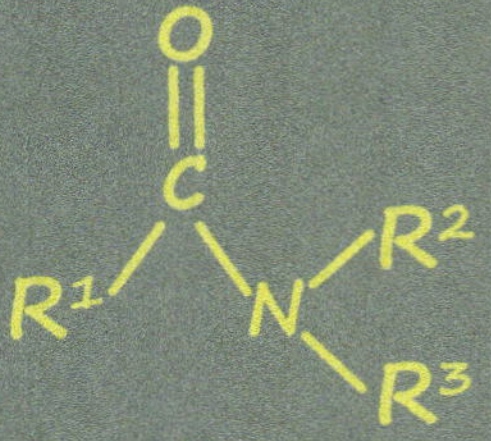

Tertiary amide

R^1, R^2, and R^3 are alkyl groups.

Some amides:

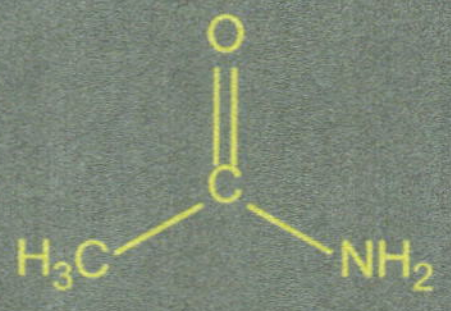

Acetamide

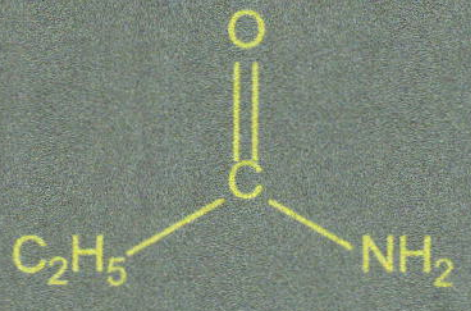

Propanamide

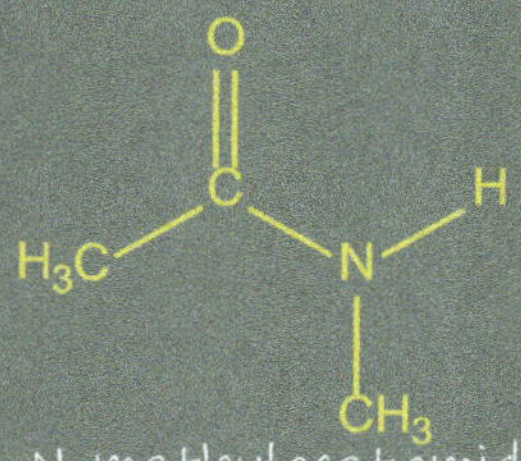

N-methylacetamide

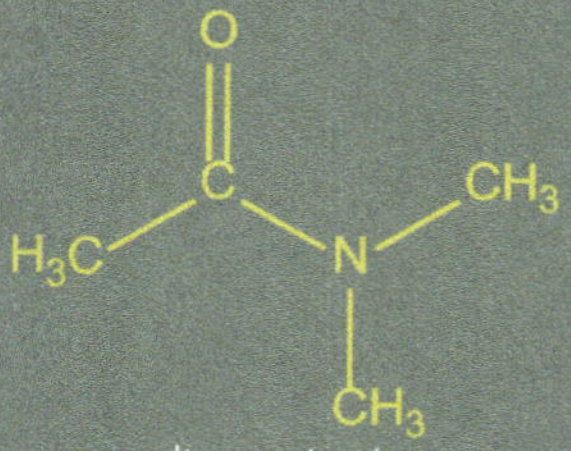

N,N-dimethylacetamide

FUNCTIONAL GROUP FOR HALIDES

Alkyl Halides

The group 17 elements in the periodic table are called halogens. These elements are fluorine, chlorine, bromine, and iodine.
Each atom of halogens is able to form one single bond with another atom.

When a halogen atom makes a covalent bond with an alkyl group, an alkyl halide is formed.
The general structure of the alkyl halides can be represented as,

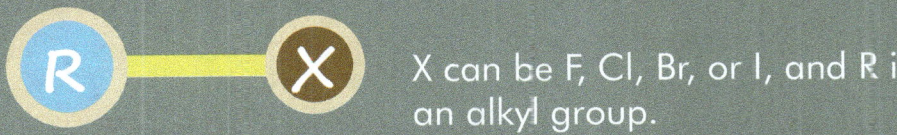

X can be F, Cl, Br, or I, and R is an alkyl group.

Some examples of alkyl halides are,

This was the short discussions about some functional groups, but the list of functional groups in organic chemistry does not end here. When you read advanced organic chemistry books, you will learn about more functional groups and their reactions.

SIMPLE ORGANIC COMPOUNDS THAT CONTAIN PHENYL RINGS.

Previously we talked about the organic molecule called benzene, which has a ring shaped structure containing six carbon atoms and six hydrogen atoms.

The shared pairs of electrons in the double bonds are delocalized due to resonance. So the carbon-carbon bonds in the benzene molecule are neither true double bonds or true single bonds, but in between these two, which is 1.5 bonds.

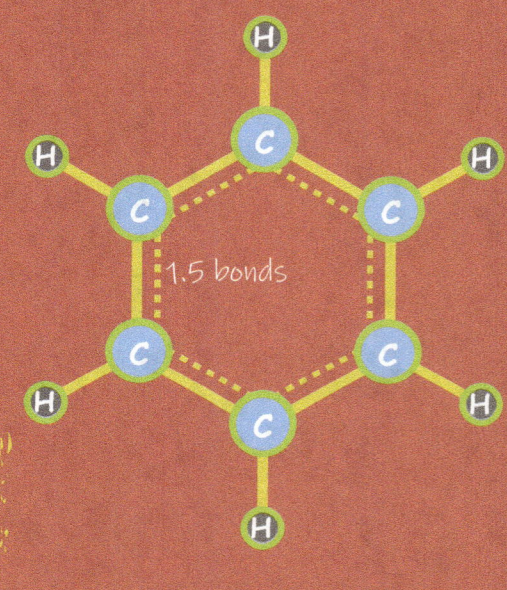

BENZENE MOLECULE

Due to that property, the benzene ring is quite stable, and it can not be broken by ordinary chemical reactions.

However, the hydrogen atoms, which are bonded to carbon atoms, can be substituted by some atoms or functional groups. Therefore, several derivatives of benzene are known to us.

The phenyl group can be represented as,

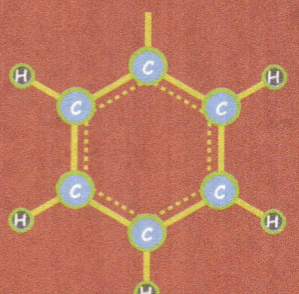

We just removed one hydrogen atom from a benzene molecule in order to create an available site for covalent bonding with a carbon atom.

A series of functional groups may attach to the corresponding carbon atom of the phenyl group, and several benzene derivatives are possible.

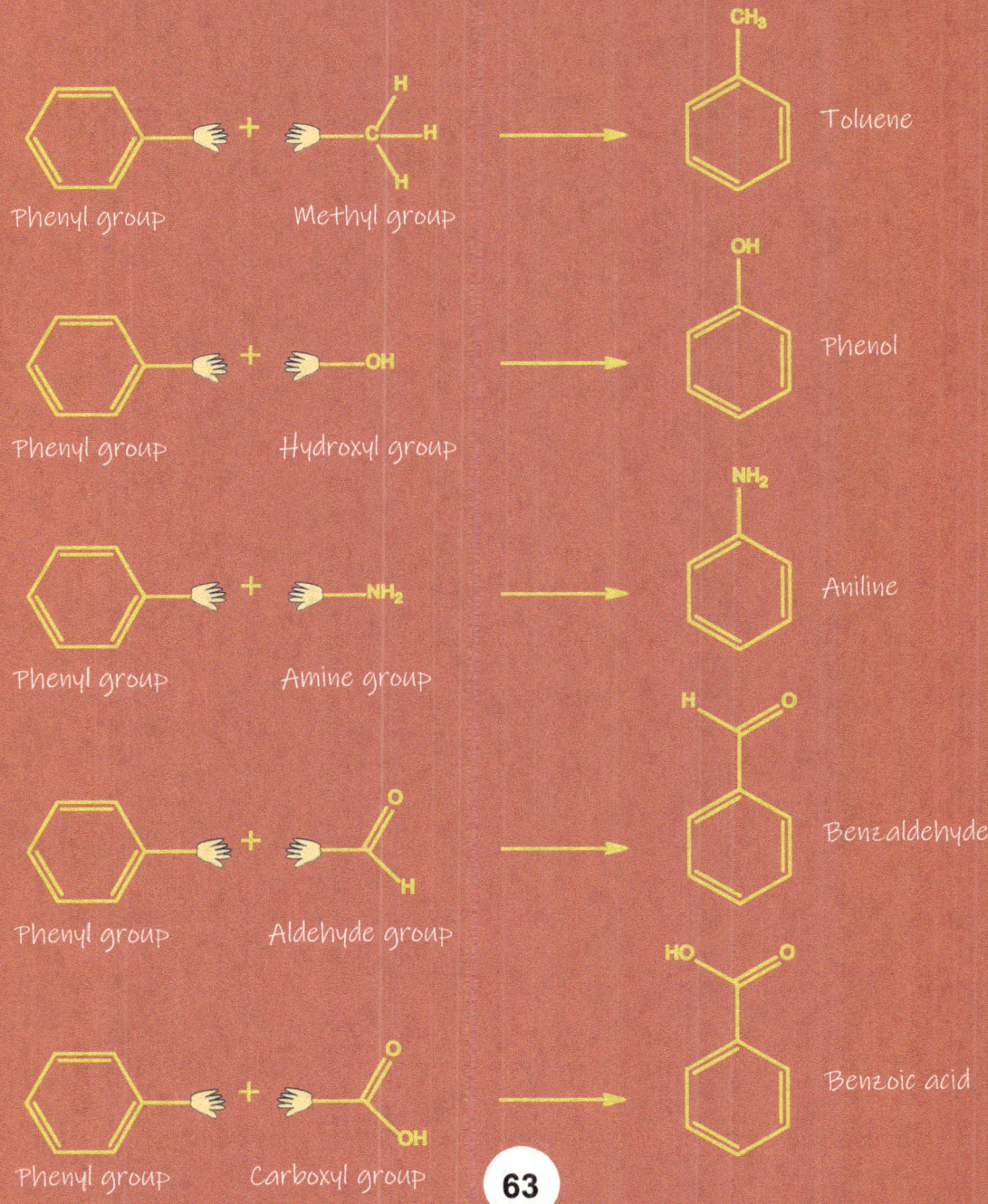

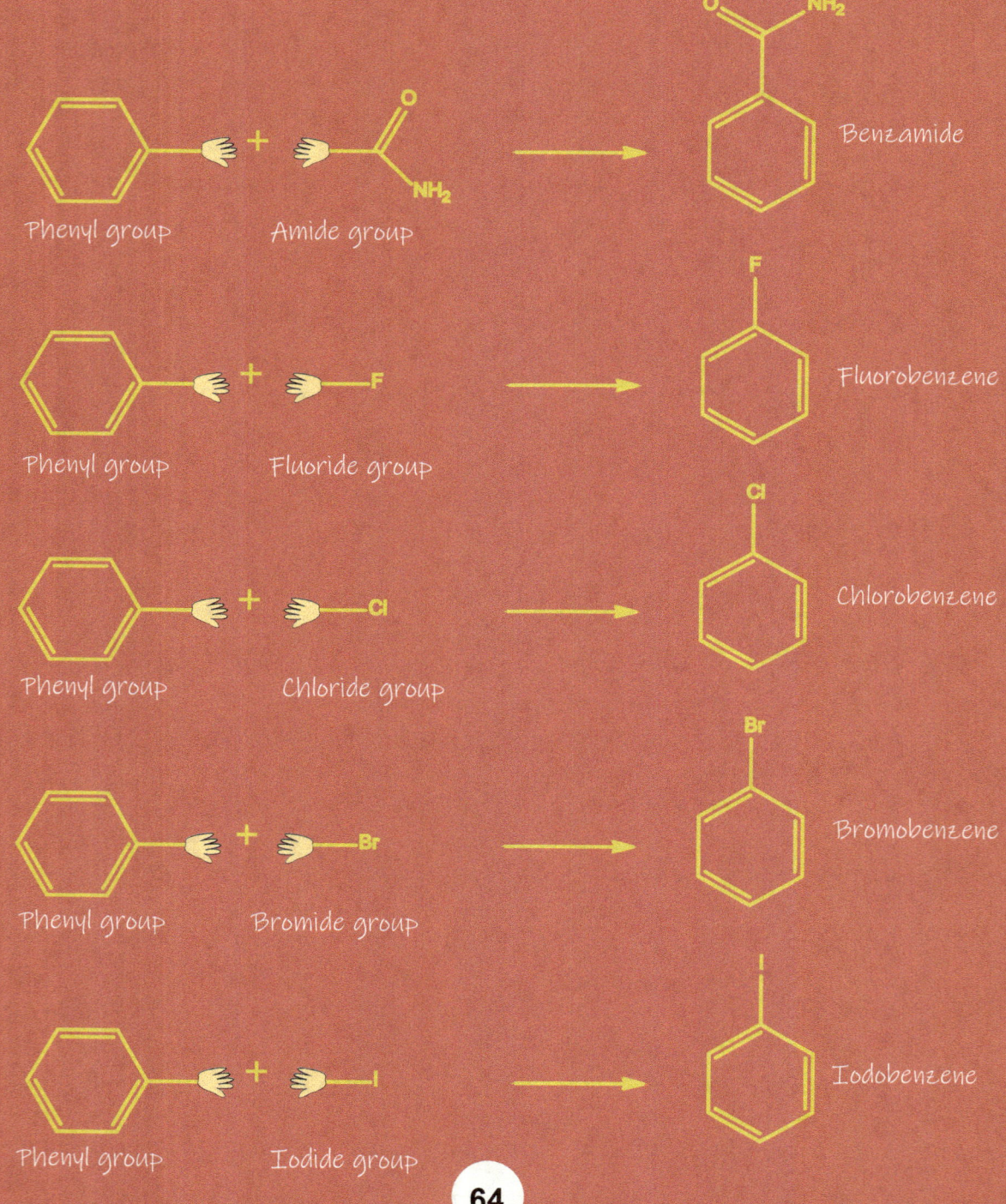

More than one functional group may attach to benzene replacing the hydrogen atoms at ortho, meta, or para positions.

Let's consider two arbitrary functional groups -X and -Y.

Suppose the functional group -X first makes a covalent bond at the available site of a phenyl group. Thus the formed benzene derivative will be look like,

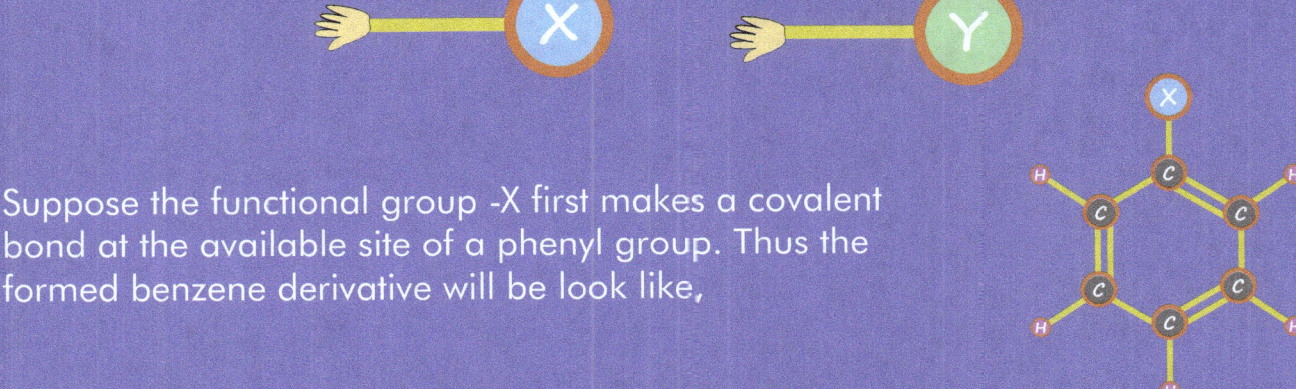

Now the second functional group -Y may make a covalent bond with any of the remaining carbon atoms replacing the corresponding hydrogen.

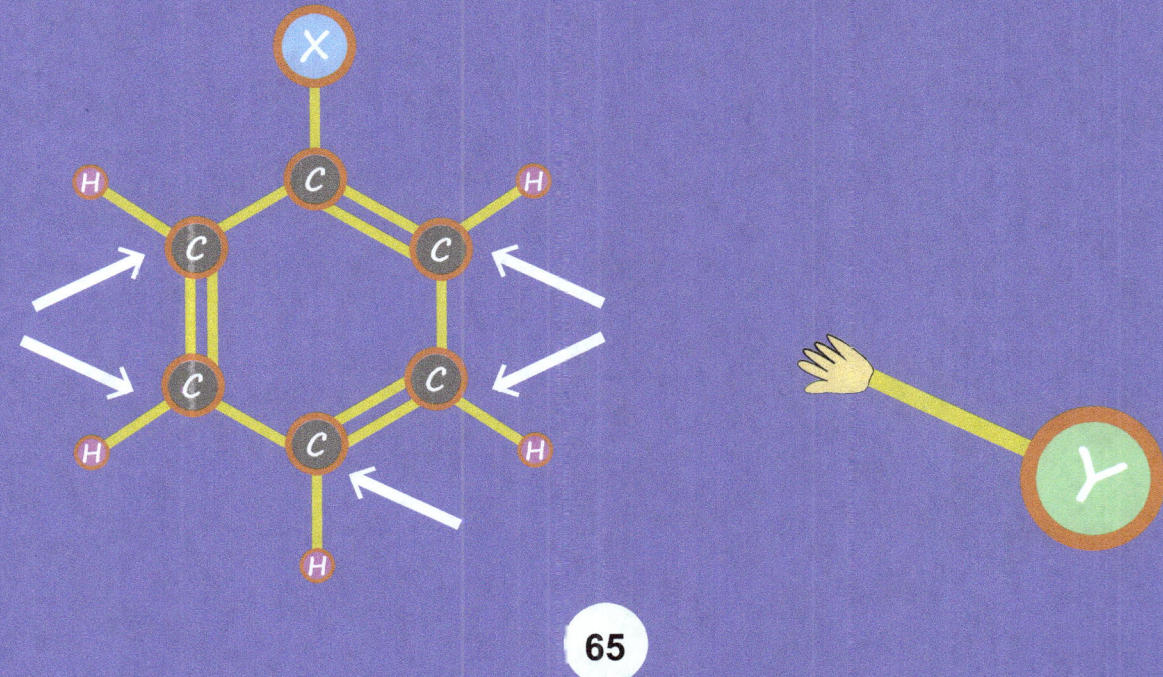

If we number the carbon atoms 1 to 6 starting from the position where the functional group -X is bonded. Therefore, the available positions for further substitution are 2, 3, 4, 5, and 6. The substituted phenyl ring will look like the following figures.

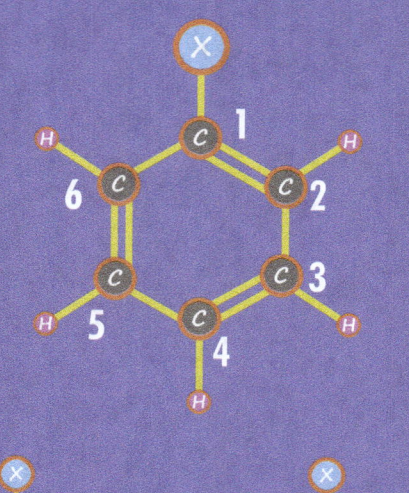

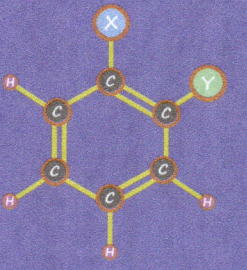 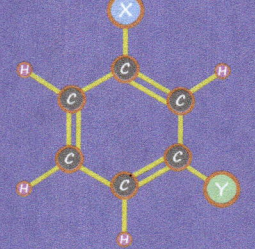 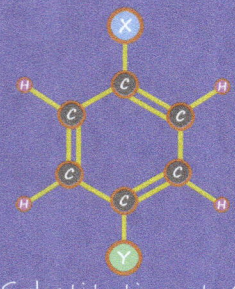 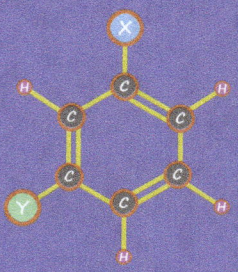 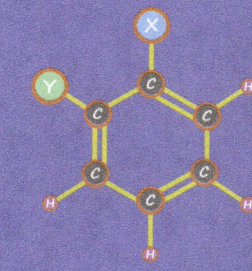

Substitution at 2 Substitution at 3 Substitution at 4 Substitution at 5 Substitution at 6

Notice that the substituted products at position 2 and 6 are identical, because if we just rotate any of these two, we get the other structure.

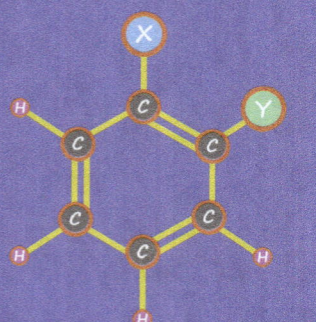 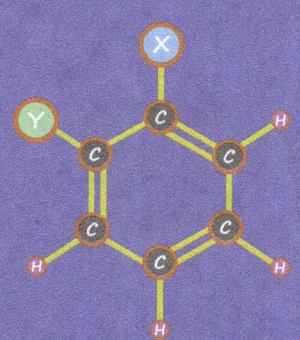

This substitution at the positions 2 or 6 is called the ortho substitution.

Ortho Substitution

The substituted products at the positions 3 and 5 are also identical, and that substitution is called the meta substitution.

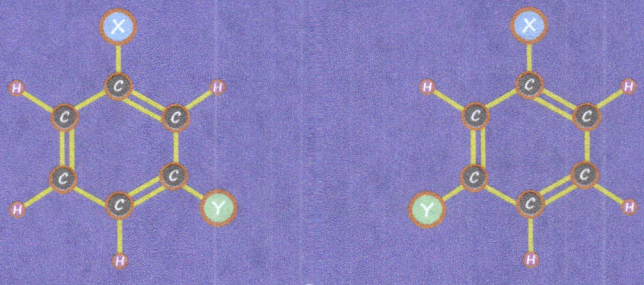

Meta Substitution

And the third possibility, which corresponds to the substitution at number 4 carbon atom, is called the para substitution.

Para Substitution

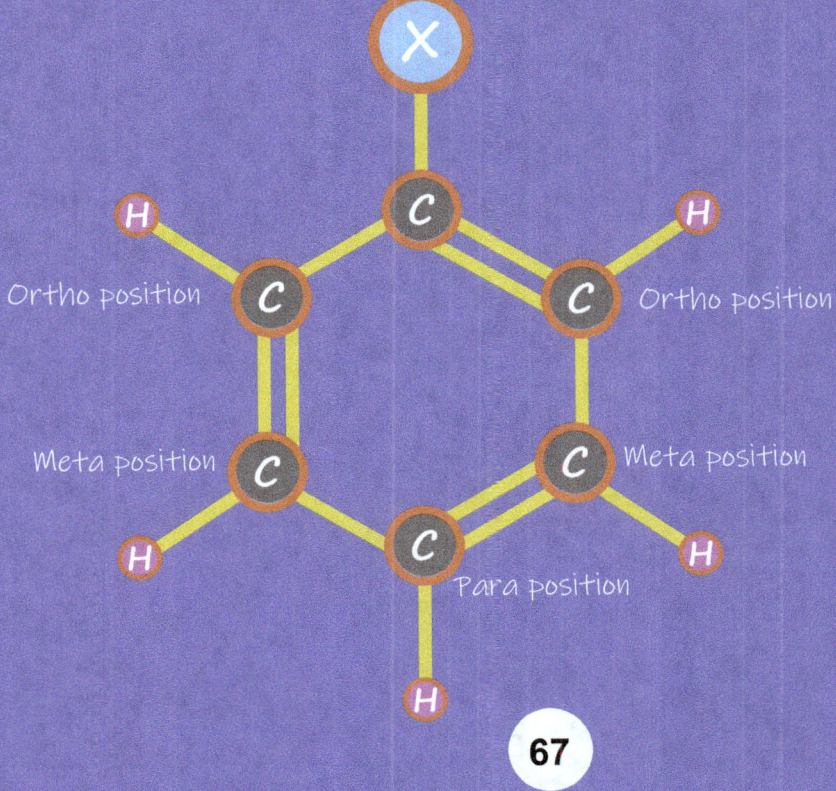

Several di-substituted, tri-substituted, and multi-substituted derivatives of benzene are known in organic chemistry.

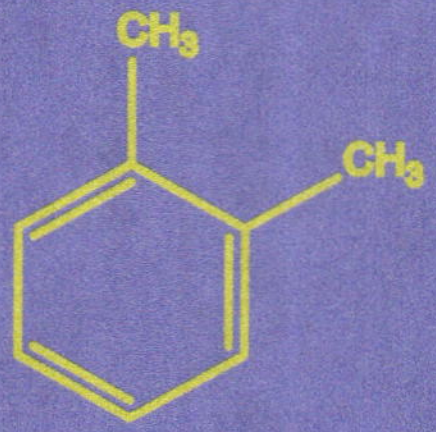

Ortho xylene

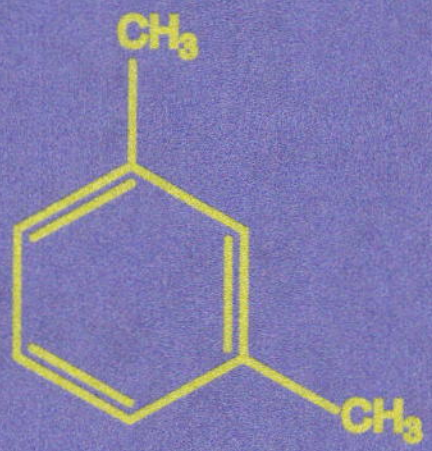

Meta xylene

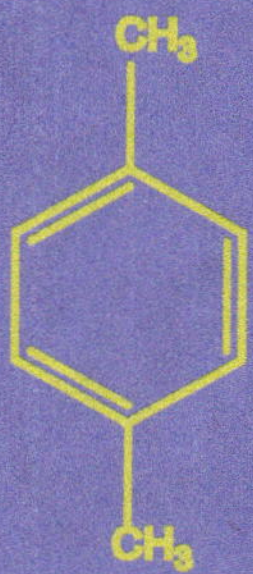

Para xylene

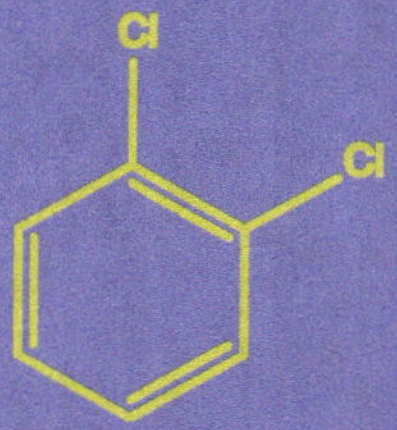

Ortho dichlorobenzene

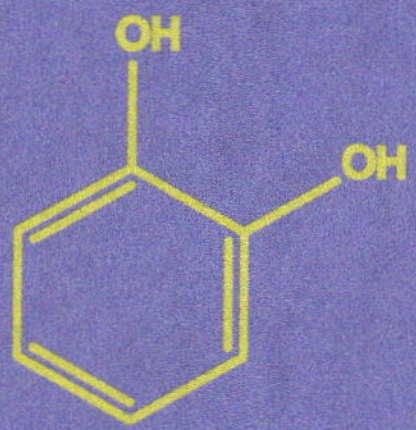

Catechol

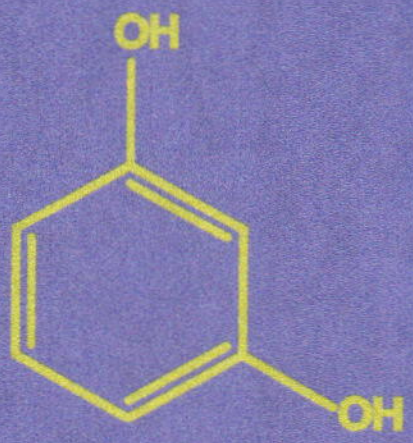

Resorcinol

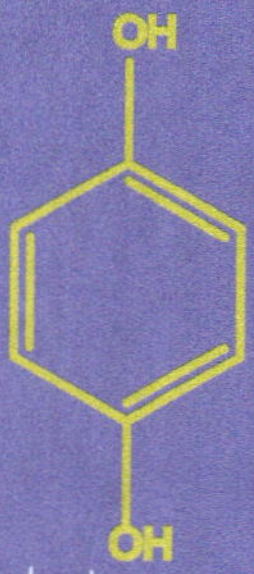

Hydroquinone

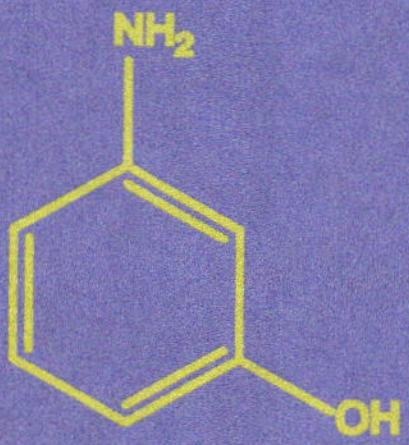

3-aminophenol

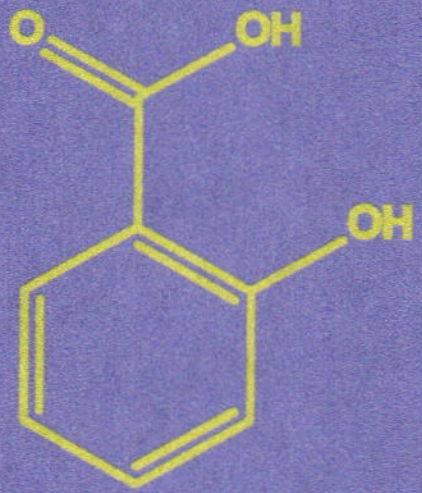

Salicylic acid

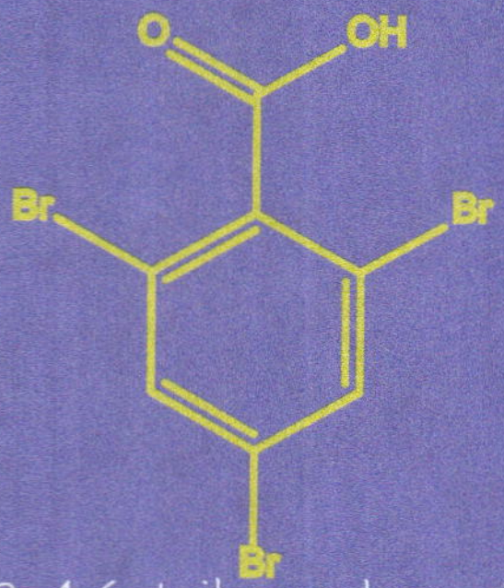

2,4,6-tribromobenzoic acid

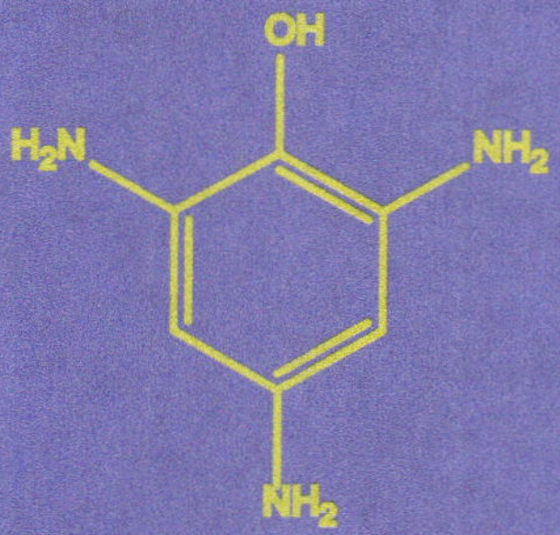

2,4,6-triaminophenol

Heterocyclic Compounds

You already know what cyclic compounds are in organic chemistry. You remember the cyclic alkane compounds known as cycloalkanes.

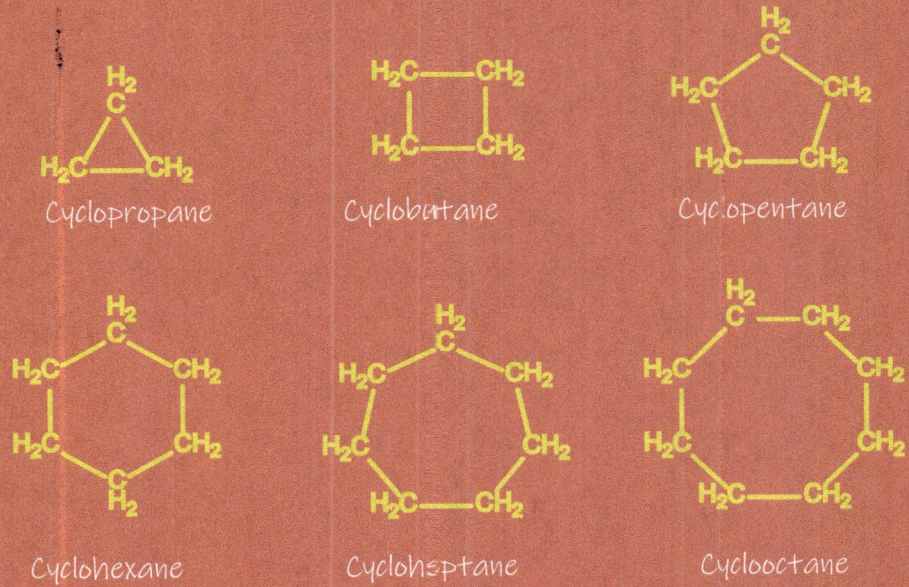

Notice here that all the atoms, which form the ring shape, are carbon atoms. But cyclic compounds can also be formed when one or more atoms on the ring are other than carbon.

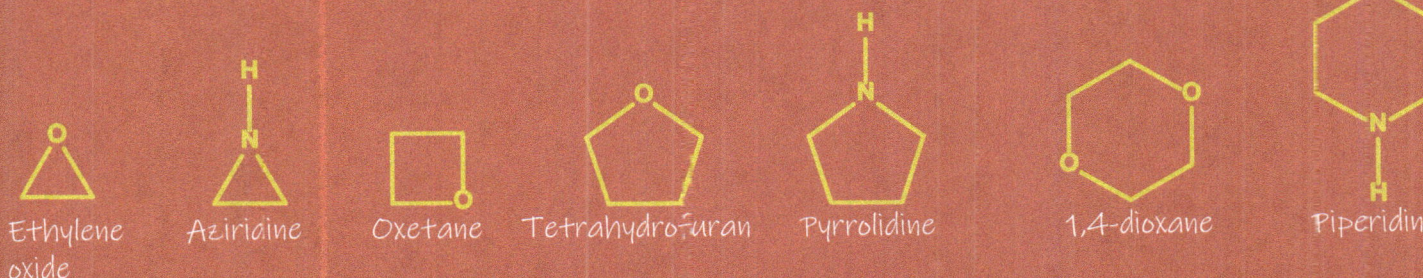

These compounds are commonly termed as heterocyclic compounds because each ring is made of at least two types of atoms.

Some heterocyclic molecules also show resonance like benzene. The examples are,

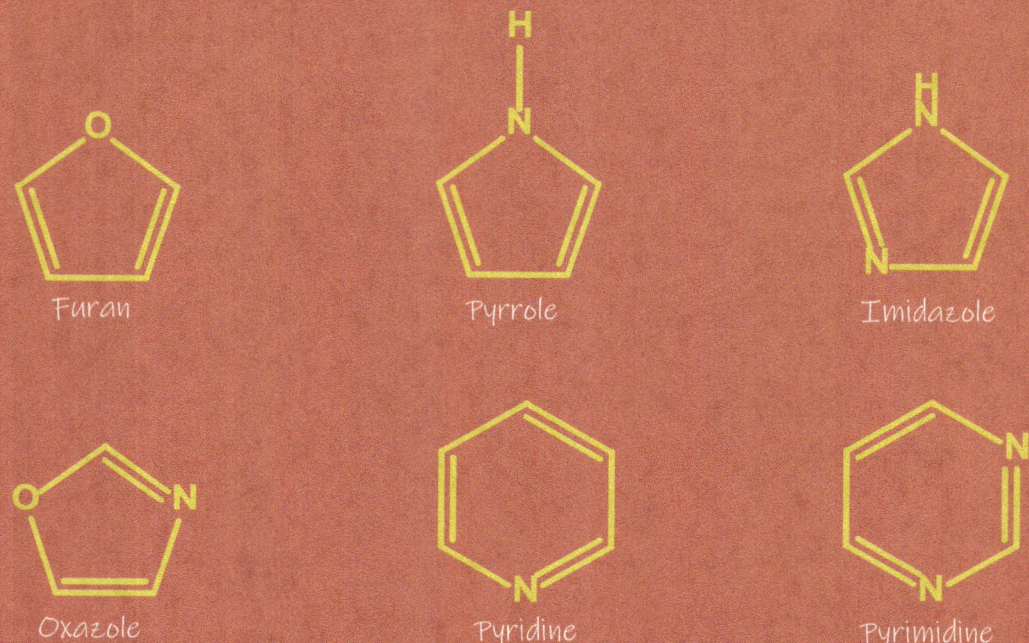

Some heterocyclic compounds contain two fused rings.

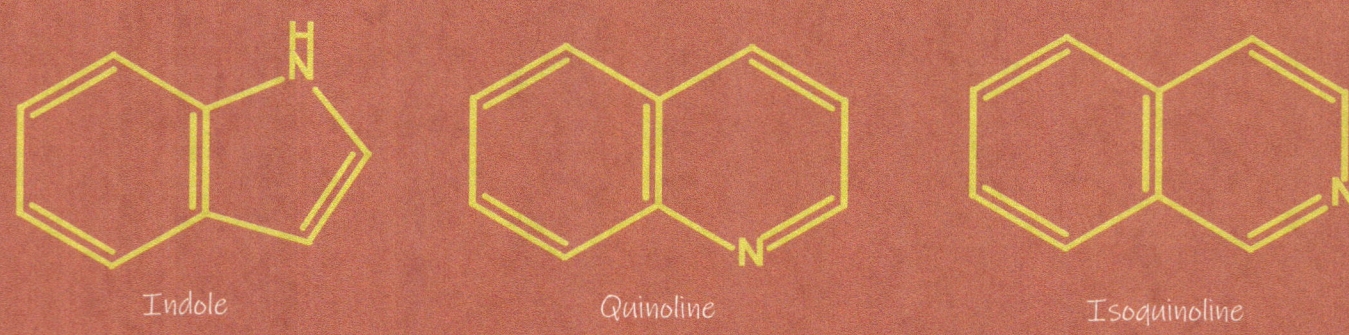

Some Organic Compounds and Where to Find Them

Apple is a popular fruit in the world. I hope you also love to eat apples. Do you know apples contain a dicarboxylic acid called malic acid?

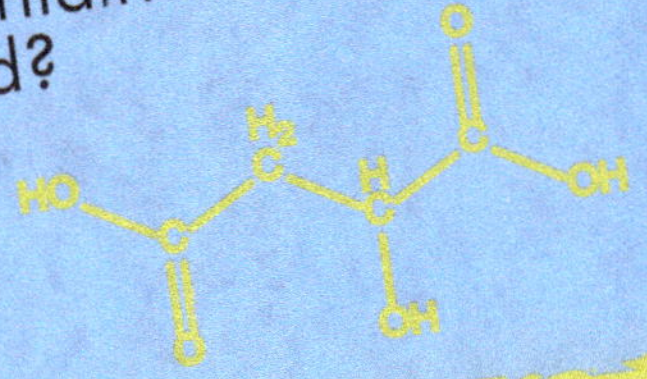

Malic acid

Tamarinds and grapes contain a dicarboxylic acid called tartaric acid.

The structure of tartaric acid looks like,

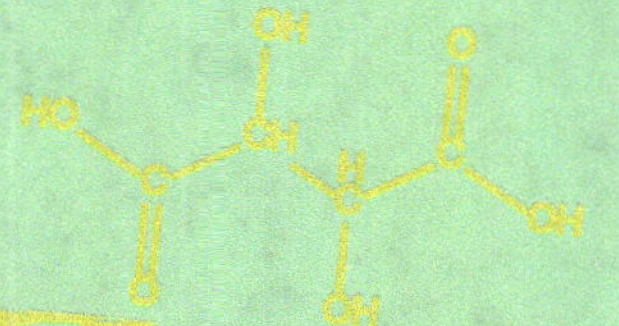

Tartaric acid

Several citrus fruits, such as limes, lemons, oranges, grape fruits etc. contain citric acid. Citric acid has three carboxylic acid groups. Citrus fruits also contain ascorbic acid, which is commonly known as Vitamin c.

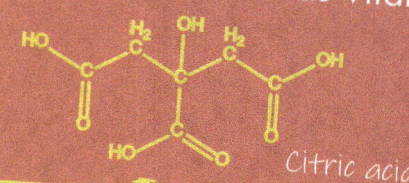

Citric acid

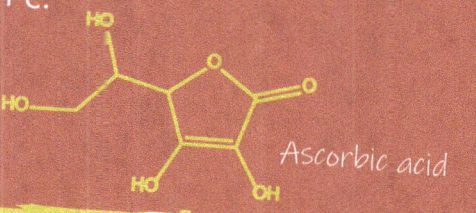

Ascorbic acid

You know curd is produced from milk by the help of Lactobacillus bacteria. During the fermentation process of milk, those bacteria produce an organic acid called lactic acid.

Curd tastes sour due to the presence of this acid.

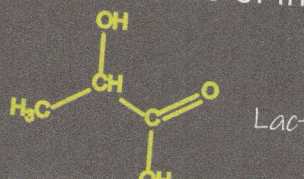

Lactic acid

Wine is produced by the fermentation of fruits and grains. During the process, ethyl alcohol is produced.

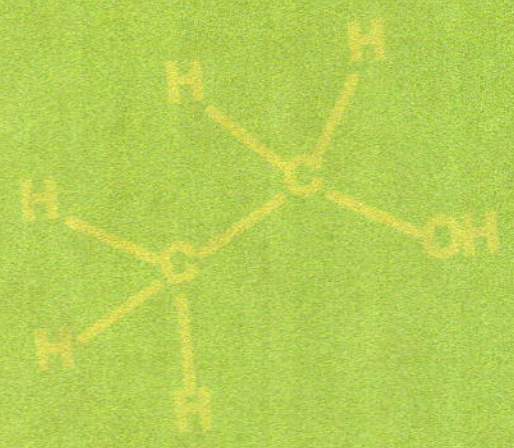

Ethyl alcohol

The organic compound called sucrose is commonly known to us as table sugar. The sucrose molecule has a large molecular structure.

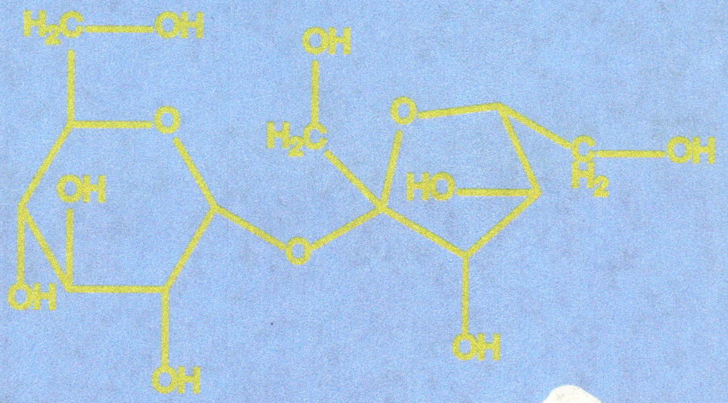

Sucrose

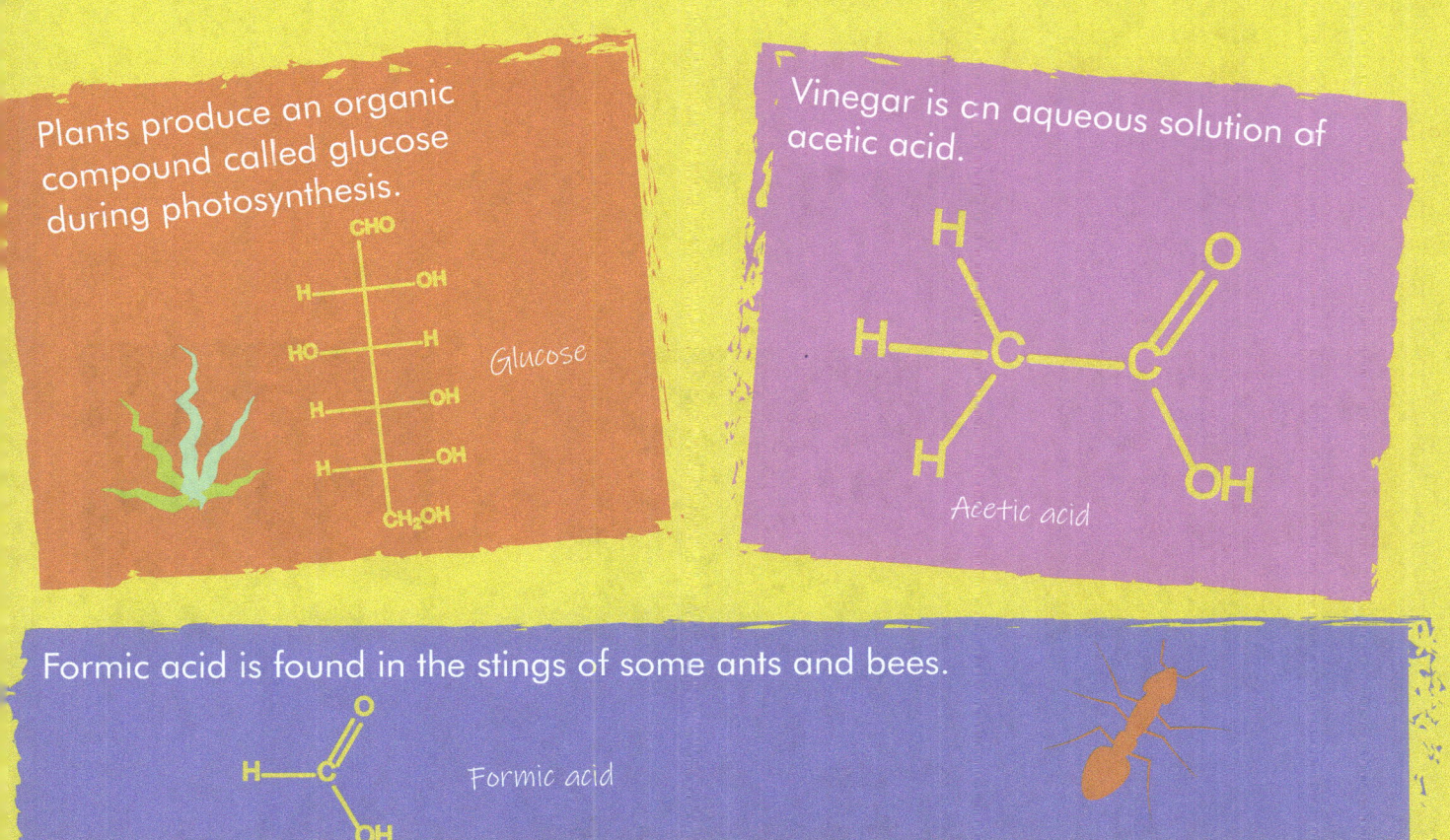

There are millions of organic compounds around us. It is quite impossible to mention all of them in this small book. However, when you finish reading this book, you realize you now know many interesting facts about organic compounds. This learning will help you in the future.

THE END

Read More Books by
JOLPIC KIDZ

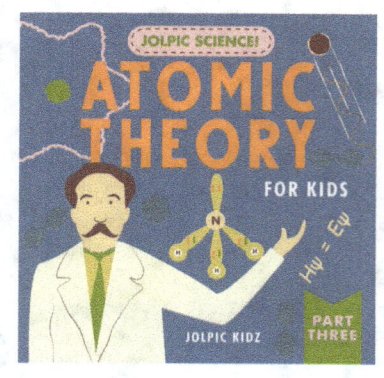

www.ingramcontent.com/pod-product-compliance
Lightning Source LLC
Chambersburg PA
CBHW082217220526
45470CB00010B/3202